SYSTEMS ANALYSIS AND DESIGN

Systems Analysis and Design

D.R. Jeffery

Department of Information Systems
University of New South Wales

M.J. Lawrence

Department of Information Systems
University of New South Wales

Prentice-Hall of Australia Pty Ltd

Prentice-Hall of Australia Pty Ltd, Sydney
Prentice-Hall International Inc., London
Prentice-Hall Canada Inc., Toronto
Prentice-Hall of India Private Ltd, New Delhi
Prentice-Hall of Japan Inc., Tokyo
Prentice-Hall of Southeast Asia Pte. Ltd, Singapore
Whitehall Books Ltd, Wellington
Prentice-Hall Inc., Englewood Cliffs, New Jersey

Typeset by Meredith Trade Lino Pty Ltd,
Burnley, Victoria.

Printed and bound in Australia by Globe Press Pty Ltd,
Brunswick, Victoria.

2 3 4 5 88 87 86 85 84

National Library of Australia
Cataloguing-in-Publication Data

Jeffery, D. Ross (David Ross), 1948-
  Systems analysis and design.

  Bibliography.
  Includes index.
  ISBN 0 7248 1181 8.

  1. Electronic data processing. 2. System analysis.
  I. Lawrence, M.J. (Michael John), 1941-   . II. Title.

001.6'1

U.S. ISBN 0-13-880261-0

CONTENTS

PREFACE

While information systems have been developed and used since time immemorial, the increasing complexity of modern organizations, together with the ready availability of inexpensive computing machinery, has concentrated attention on the process of analysing and developing systems. This has led to the emergence of a body of knowledge on how the process should desirably be undertaken. This text is intended to introduce the student to this body of knowledge: the tools, techniques and methodologies for both analysing and designing an information system. The development framework presented is based on the top-down structured approach maintaining a clear distinction between the logical and physical aspects of the system.

In designing this text to fulfil the requirements of a first course in information systems it has been assumed the student will have studied, either as a pre-requisite or a co-requisite, elementary technical aspects of computer hardware and programming. Without this exposure it would be difficult for the student to appreciate the application of computer technology to an information system.

We believe this book will suit introductory systems courses in specialized programs such as computer science and in user application disciplines such as accounting, business and librarianship. It will also serve as a handbook of design procedures for practitioners.

The approach to the subject is oriented towards practicality, with emphasis on guidelines and procedures. However, we have attempted to combine the best aspects of a 'hands-on' step-by-step guide with a conceptual treatment of the issues and principles involved.

Two case studies, one batch and one on-line, are presented in the final two chapters to illustrate the application to real-life problems of the methodologies presented in the book. The instructor may wish to set parts of these case studies for reading and class discussion as the chapters on analysis and design are being taught.

We wish to acknowledge our debt to Mrs Marie Dyer who, without complaint, typed and retyped the drafts; to our colleagues in the Information Systems Department at the University of New South Wales who contributed in many ways to the content of this book; and to Miss Annmaree Wilson for the index. We also wish to express our gratitude to our wives and children — Cas, Anna and Michael, and Sarah, Stuart and Emily — for their patience and encouragement during the time of writing this book.

D.R. JEFFERY
M.J. LAWRENCE

1 The Development of Information Systems

1.1 INTRODUCTION

Every organization has its own information system and makes use of many other organizations' information systems in order to carry on its activities. This is true whether the organization is a large public company or whether it is a group of people trying to achieve a common purpose, such as a club or society. The types of information systems kept and used by these organizations, however, would be very different in their nature and content. They would also be very different in the way in which they came into existence. In small organizations the systems are often quite informal and have evolved slowly over a lengthy period of time to fit the particular people involved. In larger organizations the information systems should satisfy the needs of a large number of people, necessitating a more formal approach to the definition of the system requirements and a more rigid application of the rules governing the operation of the information system. Despite their large size such systems may be developed quite quickly (6 to 12 months) using large teams of people.

An information system is defined as a means of providing information in such a way that it is of use to the recipient. A very broad view of information is taken to include not only the provision of traditional management information (decision-input information), but also the processing of data associated with the routine operation of business systems, e.g. payroll processing, debtors, etc.

In the case of a small club, a simple list of its members and the details of any annual fees outstanding might provide a very effective information system for one aspect of the club's activities. This list could provide not only the details of any financial indebtedness to the club, but also details of the members' addresses and telephone numbers which might provide the basis for all club activities. Along with the club's monthly bank statements and cheque butts, this may be the only formal system within the club. This is not to say that it is the only information that is relevant to the club, but rather the only information for which a formal mechanism may be developed for capture, storage and dissemination.

In a large public company a simple information system would be completely inadequate. The requirements of business and legislation are such that many information systems need to be developed to track the organization's activities and to provide a basis for its future activities. In one sense, the information systems of an organization are concerned with the past, in terms of providing a record of what has gone on in the organization over a period of time; the present, in terms of conveying the status of the organization at this point in time; and the future, in terms of the information that can be provided to assist in determining the best strategy for future action.

1.2   DEVELOPING SYSTEMS

The manner in which a new information system is developed depends on a number of factors, such as the size of the system, the complexity of the system, and the characteristics of the system's user group. Information systems literature contains many references to the 'user'. This term covers a very broad class of people within the organizational structure, ranging from (say) a machine operator up to the managing director. They all participate in information systems in the roles of providing data input, receiving routine data output, or making decisions based on the information supplied by a system. Consequently, the skills and knowledge base of a system's user group can vary considerably, particularly from one system to another. For this reason the systems analyst needs to be effective in communicating with a very broad range of people.

The development approach taken throughout this book follows a fairly traditional structured methodology. This approach is particularly applicable when a system is fairly large, complex, or in some way stretching the capabilities of the organization. When a task is very simple, or required very urgently, a more expedient approach may be more suitable.

1.3   THE SYSTEMS DESIGN TASK

Developing large information systems is a professional activity, requiring both judgement and creativity for a successful result. In meeting the requirements of the user, the analyst combines current technology with the skills of the user and the skills and expertise of the other members of the system's development group. This process requires considerable creativity. There are many possible ways of structuring a system, each with its own particular strengths and weaknesses. To assess the tradeoffs associated with each particular alternative, the professional judgement of the systems analyst is required.

1.4   THE TRADITIONAL LIFECYCLE

In order to provide a structure for the development of information systems, most organizations follow what is called a system lifecycle approach. This lifecycle is in many cases a list of discrete stages through which the project proceeds as it is developed:

Stage 1   Requirements specification
Stage 2   Feasibility study
Stage 3   Logical design
Stage 4   Physical design
Stage 5   Programming
Stage 6   Implementation
Stage 7   Post-implementation review

This lifecycle encompasses the two important characteristics of top-down development and structured design. These concepts are discussed in Chapter 7, but an overview would be helpful at this stage. The lifecycle and its associated methodology provide a step-by-step approach to development. At each stage of the design more detail is added until the physical specification is complete in Stage 4. A distinction is drawn in the lifecycle between logical and physical systems. A logical system is concerned with 'what do we want the system to do' and a physical system with 'how will it be done'.

Within each of the lifecycle stages there exists a set of activities to be carried out, such that the documentation might take the form of a manual for the construction of systems, not unlike a large construction project. The tasks within each stage have to be defined and completion dates set so that control over each stage and hence over the whole project can be achieved. Thus the lifecycle provides not only a framework for developing a system, but also a basis for controlling each stage of the project.

Requirements specification. This stage is concerned with reducing uncertainty surrounding the scope of work to be carried out. This usually means looking at the system's target area and gaining an understanding of its environment, identifying any pressures for change on the current system, and looking for areas in which improved performance which may be possible if new systems were implemented. The amount of effort involved in this stage will vary enormously depending on factors such as the complexity of the environment, the risk associated with a change, or the technology available to realize a particular advantage. Therefore the requirements specification task needs skills such that opportunities can be perceived, knowledge of the organization such that the environment under study can be appreciated, and knowledge of available technology such that any advantage can be gained through the use of suitable technology. These skills are very different from those required from (say) a programmer and indicate some of the skills that are needed in a team which is responsible for the development and implementation of an information system.

Feasibility study. This stage focuses on analyzing the current system and proposing alternatives which meet the requirements of the user. It is here that most of the systems analysis work is usually carried out. The aim is to explore the alternatives available to such a depth that it is possible to determine which option is the most desirable from both a technical and an economic viewpoint. Not only is an analysis of the current system carried out, but quite often some aspects of the design of the new system. It must be remembered that the feasibility study should only go as far in the analysis and design as is necessary to establish feasibility. It is a most important stage in the lifecycle because it sets the foundations for all future work and dictates the direction that the design will pursue.

Logical design. In this stage the designer sets about the task of producing a complete logical description of the new system. This entails a description of the data and data flows in the system, the inputs and outputs of the system, the control and security procedures that are necessary, and the logical operation of the system in the user environment. At this stage the analyst determines what the computer system will be doing, but does not provide all of the necessary details to completely specify how the system will operate.

Physical design. In this stage the designer fills in detail, or provides meat for the skeleton produced in the logical design. This requires the complete specification of all inputs and outputs, files or data base, security and control procedures, and the operation of the system in the user environment.

Programming. Typically the physical design must now be converted to a set of program specifications. These specifications are then given to programmers to design and code the system. A critical part of the programming task is the testing of programs both at an individual level and at the system level — where they all have to function together to provide the complete system.

Implementation. This stage aims to provide an operational system for the user. Usually this involves the installation and testing of the complete system in the user environment. Frequently the new system is run in parallel with the old in order to provide a thorough test of the new system by reconciling its output with that of the old. In other cases this parallel running is not possible and other procedures are used to verify the output of the new system.

Post-implementation review. After the system has been operating for 6 to 12 months a review may be undertaken to determine whether it is successfully meeting the requirements originally set. A review of the benefits being achieved through the use of the system may also be undertaken to provide feedback on the estimation procedures that were used earlier in the system's lifecycle.

1.5   INFORMATION SYSTEMS COMPONENTS

Figure 1–1 shows a generalized model of an information system. In this model we can identify the major components of an information system. These are:

- Data input
- Input error correction
- File update
- Reporting
- Inquiry processing

Any information system will be composed of these five basic parts, arranged in various ways to satisfy user needs.

Data input. Transaction data (e.g. orders received, products sold) are entered into the system.

Correct errors. All data input has potential for errors, and must be checked and corrected where necessary before information processing can proceed.

Update files. The transaction data (now stored in a machine readable form) are used to update the company files (e.g. stock files, employee files).

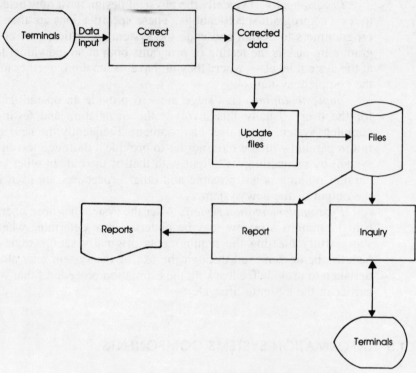

Figure 1–1 Information system model

Report and inquiry. Both of these components are concerned with accessing stored data to provide information for use in the organization. An inquiry may lead to either the printing of a report or the display of information on a visual display unit (VDU). The types of reports provided for management can be divided into four basic types:

- Scheduled reports
- Predictive reports
- Demand reports
- Exception reports

Scheduled reports can be further subdivided into those reports covering activities and those covering status. An example of an activity report would be a production report showing quantities of goods manufactured during a period by different sections of the organization. A status report, by way of contrast, might show the balance of work in process at each manufacturing point at a particular point in time.

Predictive reports show the expected or possible results of particular courses of action. A budget is one form of predictive report, while the results of a model of some part of the organization is another.

Demand reports are those which are requested by management. These can be subdivided into preformatted and *ad hoc* reports. A pre-formatted demand report is one in which the layout or format is pre-determined and the report supplied to management when requested. An ad hoc report is one in which the format of the report is not known until the request for information is supplied by management.

Exception reports, as the name suggests, show only those items which do not satisfy a given rule. This type of reporting is very important because it concentrates management attention on important items and does not show items where no management action is necessary. An example of this is in an accounts receivable system, where management is interested only in slow payment accounts, or debts above certain values. It should be noted that scheduled reports, predictive reports and demand reports can all be prepared on an exception basis.

1.6 DATA AND PROCESS DESIGN

In the lifecycle section the development of systems was presented as a layered set of stages. Another way of looking at the development task is to consider a further sub-division of each of the lifecycle stages into three segments:

- The data segment
- The process segment
- The context segment

These three segment types are shown in Figure 1–1: the data (files, corrected data), the process (update files, correct errors, report, inquiry) and the context (people using the terminals for data input or inquiries, or people using the printed reports).

Throughout the development of a system the designer is deciding on the database or set of files which the information system will use, the processes which are carried out within the information system, and the procedures which will be used in the organization in order to use the information system. In this way the designer undertakes a database design or data analysis, a process design, and also an organizational design set of activities. These three activities take place in all stages of the analysis and design of the new system. This concept is illustated in Figure 1–2.

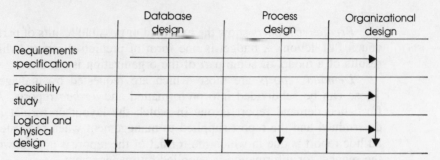

	Database design	Process design	Organizational design
Requirements specification			
Feasibility study			
Logical and physical design			

Figure 1—2   The three elements of design

1.7   ALTERNATIVE LIFECYCLES

Over the last twenty years, since the introduction of programming languages such as COBOL, BASIC and FORTRAN, there has been little change in system development productivity. This has motivated the development of:

- Very high-level languages
- Automated approaches to design
- Prototyping of systems to reduce subsequent maintenance costs

With the development of much more powerful systems software (designed to help the system builder), it has become possible to approach the development of certain classes of systems in a slightly different way. For example if a system is being developed which uses an existing database, and a less procedural (or very high level) language is available, quite often an evolutionary approach has been used to advantage. The nature of the evolutionary approach to software development is shown in Figure 1–3. After the requirements specification stage, a prototype for the system is designed and constructed, usually using the power of the very high level language. This results in a system shell which can be provided for the users to gain experience and to evaluate. As a result of this evaluation there are typically some aspects of the system's operation which are unsatisfactory to the user, and consequently modifications are identified and the prototype updated. This prototype is then evaluated again by the users until such time as a satisfactory system is established. This is not necessarily a complete system, but only a prototype which specifies how that system will look to the users. It may not contain all of the control and security procedures of the eventual system, nor (say) the full database specification.

It is still early days in the development of very high level languages and consequently in the assessment of the best systems lifecycle to use with these languages.

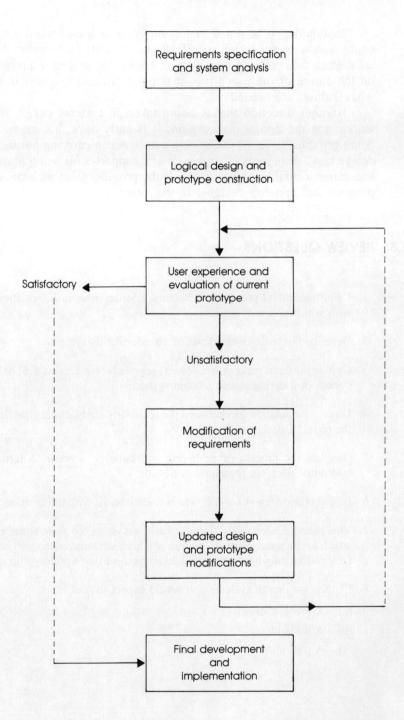

Figure 1—3　Evolutionary software development model

It should also be noted that prototyping is useful when a system is being developed for which specifications are difficult to write. This may arise when the system is very new, or the system will have a large impact on the nature of the user's task making it difficult to guess in advance what features are needed.

Another direction that is being taken in systems design is that of automating the design task. Again, it is early days, but many aids are being provided to assist the system's designer in carrying out the system design task. Aids such as system design support-tools using graphics are becoming available and will no doubt provide great assistance to the program and system's designer in the future.

1.8   REVIEW QUESTIONS

1. Using the model provided in Section 1.5, describe an information system with which you are familiar.

2. Describe the basic components of an information system.

3. Outline the data, process and context segments (see Section 1.6) of a payroll system or a savings bank account system.

4. Discuss the relationship between the feasibility study, the logical design and the physical design.

5. Describe the process of designing and building a house in terms of the traditional lifecycle presented in Section 1.4.

6. Discuss the different ways in which users interact with information systems.

7. You have decided to design and build a system for your home computer which keeps a record of all movies you have seen and books you have read. Describe broadly how you would go about this using a prototyping approach.

8. List the major sub-systems you would expect to find in:
   (a)   A grocery store
   (b)   A trade union
   (c)   A coal mine

2 System Representation Tools

2.1 INTRODUCTION

As seen in Chapter 1, the process of developing computer-based information systems consists of a number of stages within a system lifecycle. At each of these stages different tools are applicable to assist in the system's development process and to document the work done. For example, when analyzing an existing physical system it may be advantageous to represent the flow of documents through a functional flow diagram, or when representing an existing computer-based system a set of system flowcharts may be best. Over the years many tools have been developed to assist systems analysts in their tasks. This chapter looks at the following tools:

- System flowcharts
- Functional flow diagram
- Data flow diagrams
- Hierarchy charts
- Input process output diagrams

Other tools have been developed, but this subset represents the tools more commonly used today.

These tools are largely used for charting system characteristics. As such they provide two functions: they assist the analyst while he is attempting to gather his thoughts, either on the existing system or on the proposed new system; and they provide a hard copy record of the work carried out for later reference. When developing systems there is always the need both to carry out the work and to communicate it.

2.2 FLOWCHARTS

Flowcharts are schematic representations of the processes and logic involved in an information handling activity. The flowchart illustrates each step required in the activity by using symbols which depict the nature of the activities or steps to be carried out. The flowchart is one of the oldest systems tools — a set of standards were devised in 1963 by the American Standards Association Committee on Computers and Information Processing. These symbols were revised in 1966 and have been widely accepted as a standard. The reasons for the wide use of flowcharting as a tool for system development are its simplicity and the benefits obtained through its use. These benefits include:

- *Relationships*. The flowchart provides a very effective, clear and concise description of logical procedures. When compared with text the benefits of a graphical representation are obvious, particularly for complex logic.
- *Analysis*. The clarity of expression afforded by the model allows the user to pinpoint problems or to highlight particular areas of the system.
- *Communication*. Provided the symbols used conform to a standard, the flowchart provides a convenient tool for communicating problem resolution logic, and for documenting that logic.

There are two basic types of flowcharts: the system flowchart and the program (logic) flowchart. The relationship between these two types is shown in Figure 2–1.

The system flowchart provides a representation of a system, or part of a system, whereas the program flowchart shows the logic of a process in that system. In this figure part of an inventory system is represented in the system flowchart, and the program flowchart shows the logic of the 'print order list' process. The following sections cover system flowcharts which are concerned with the inputs, processes and outputs of systems. The interconnecting lines describe the flow of information to and from a process. The boxes represent the records, files, databases, processes and equipment that go to make up the physical system.

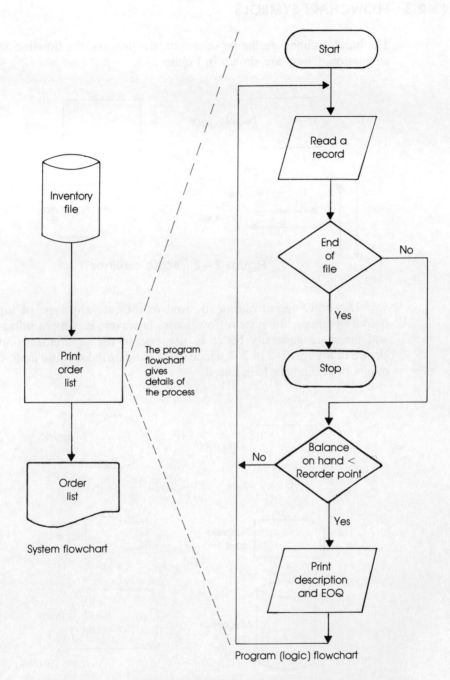

Inventory
file

Print
order
list

The program
flowchart
gives
details of
the process

Order
list

System flowchart

Start

Read a
record

End
of
file

No

Yes

Stop

Balance
on hand <
Reorder point

No

Yes

Print
description
and EOQ

Program (logic) flowchart

**Figure 2—1   Relationship between the system flowchart
and the program (logic) flowchart**

2.3 FLOWCHART SYMBOLS

The basic outlines are the input-output, the process, the flowline and the annotation. These are shown in Figure 2–2.

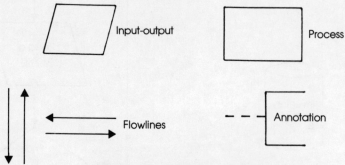

Figure 2—2   Basic outlines

The input-output outline is used to indicate any type of input or output operation. In system flowcharts, however, it is more informative and therefore generally better to use one of the 'specialized outlines' shown in Figures 2–3 or 2–4, since these indicate the specific input-output media or equipment being used.

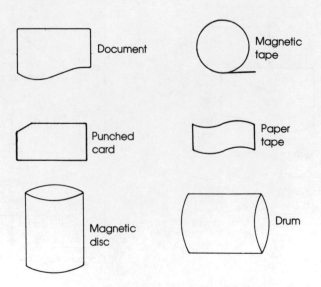

Figure 2—3   Specialized outlines for media

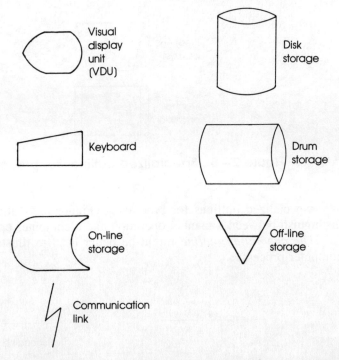

Figures 2—4 Specialized outlines for equipment

Media outlines (Figure 2–3) represent the physical media on which the data is stored. The document symbol represents a paper document which may be input to a system or output from a system. The other five symbols represent various forms of media which can be used to store data in a binary form.

Equipment outlines (Figure 2–4) allow distinction to be made between on-line and off-line storage and also between a visual display unit (VDU) or another form of key device for input or output.

The process outline is used to indicate logic operations, data movement and data transformation. In system flowcharts it usually indicates a computer process such as:

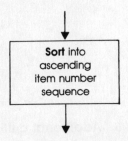

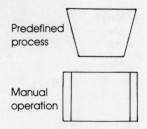

Figure 2—5   Specialized outlines for processing

With specialized outlines for processing (Figure 2–5), it is possible to distinguish between a manual operation and a computer operation.

The annotation outline might be better used to illustrate the same computer process.

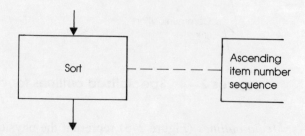

Flowlines connect successive outlines in order to show the flow of information from one activity to another. When drawing flowcharts it is normal for the flow to be represented from top to bottom and left to right.

An additional outline which is used to make system flowcharts more readable is the connector symbol. This symbol allows the chart to be broken into pieces to fit conveniently on a page and also to show exits from and entry to a particular segment of the chart. The connector symbol is shown in Figure 2–6.

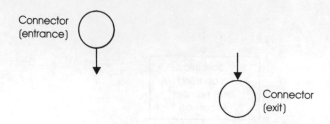

Figure 2—6   Additional outlines

Where it simplifies the flowchart or provides additional information to the reader multiple outlines should be used. With these it is possible to indicate (Figure 2–7) the receivers of the document or the fact that the system is dealing with (say) a number of files on the one physical disk device.

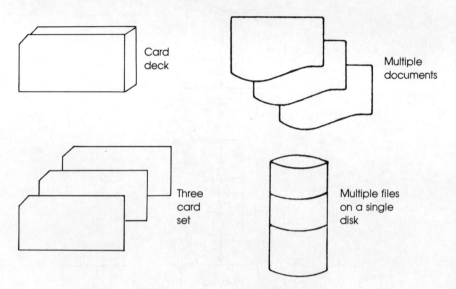

Card deck

Multiple documents

Three card set

Multiple files on a single disk

Figure 2—7   Multiple outlines

2.4   SYSTEM FLOWCHART EXAMPLE

Figure 2–8 shows the logic of part of an order entry system in which customer orders are used as input to prepare the invoice and shipping documents, and to update the debtors and inventory files with sales details. In this simplified system a batch of orders is placed into the system. At stage 1, control totals, such as the number of orders in the batch, are taken out to ensure that all transactions entering the system are processed. In stage 2, these documents are converted into a machine readable form (i.e. magnetic disk) and in stage 3 an edit is performed and the transactions written again to a new transaction file. The edit procedure checks that the transaction data appear to be valid. This may be done by ensuring that the quantity ordered is within a reasonable range, that the customer placing the order exists on the debtors file, and that the item ordered exists on the inventory file. Stage 4 carries out the update of the debtors and inventory files, prints the invoice and shipping documents, and calculates the control totals which can then be balanced with the controls established in stage 3.

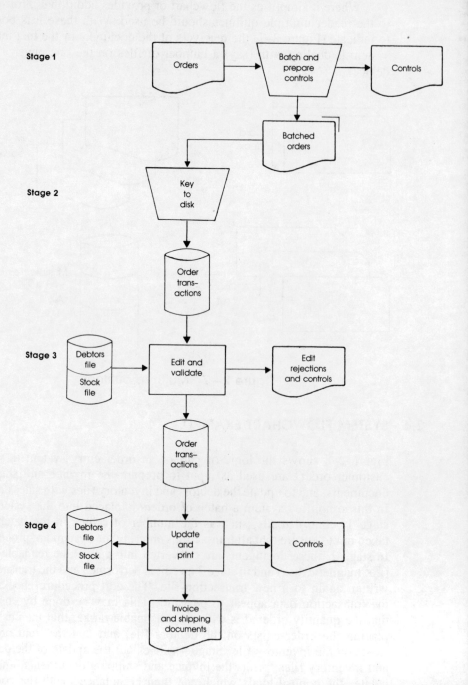

Figure 2—8   Order entry system flowchart

2.5 SYSTEM FLOWCHART ELEMENTS

It is possible to identify a number of different types of system flowchart segments. These correspond to different functions that are typically performed in systems. These functions are:

- Data entry
- Edit
- Update
- Report
- Rearrangement

At this level each flowchart is describing one particular process. Figure 2–9 shows the process of capturing data in machine readable form

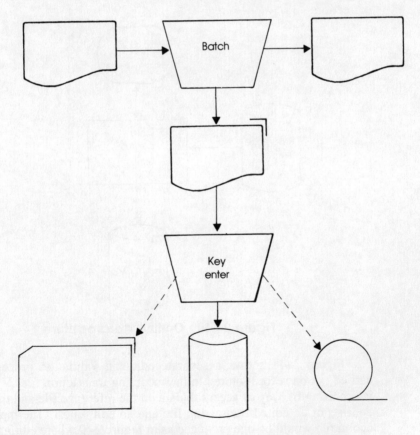

Figure 2–9   Off-line batch data capture

from documents in a batch environment. The process involves first batch-ing and establishing batch controls of the documents, and second keying the documents using appropriate equipment — either onto punched card, magnetic tape or more usually magnetic disk. Both processes here are manual and therefore describe an environment which does not involve the use of a computer program.

Figure 2–10 shows a different data capture environment in which data is entered via a VDU and a computer-based process is used to write the records on to a transaction file stored on magnetic disk. Because a computer process is involved, it may also be decided that reference files could be on-line in order to carry out validation checks against the data being entered from the screen. In the figure the arrow connecting the process with the VDU symbol is bi-directional, indicating that data is being transmitted in both directions. This might take the form of the record being transmitted from the screen to the program and (say) error diagnostics being transmitted from the program to the screen for operator correction.

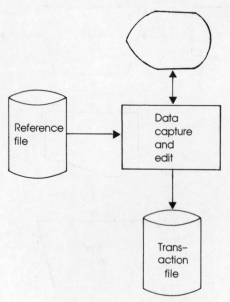

Figure 2—10   On-line data capture

Figure 2–11 shows a separate edit and validation process being carried out on records stored in the incoming transaction file. Validation is occurring by way of access to data on the reference file and the output consists of an edited transaction file and an edit report. The input trans-action file would be one produced as in Figure 2–9 where editing has not taken place.

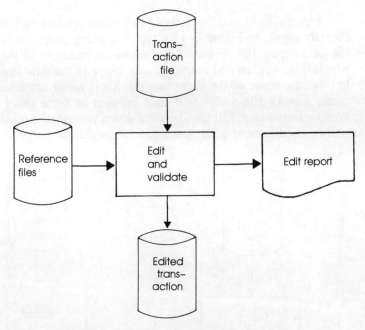

Figure 2—11   Batch edit process

An update process is represented in the flowchart of Figure 2–12. In this case a transaction file is being used as input to update records on the master file. The output of the process is the updated records on the master file and a printed update report.

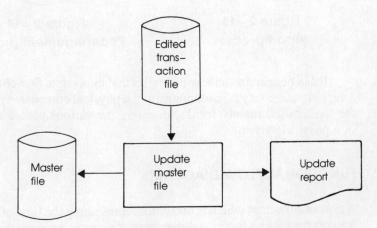

Figure 2—12   Batch update process

Figures 2–13 and 2–14 show two other processing forms in which files are input. In Figure 2–13 a report is being produced from a master file or a report file. Figure 2–14 shows an example of the situation in which it is necessary to rearrange in some way the data stored on a file. In this case some of the data from the file is being extracted in order to create a work file which will then be used at some other point in the system. For example, it may be a cut-down version of a master file which will be used later as a reference file for validation purposes.

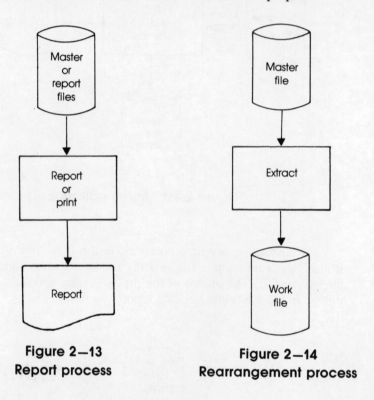

Figure 2–13
Report process

Figure 2–14
Rearrangement process

It can be seen from these examples that the system flowchart provides a very effective way of communicating a physical computer system. Thus the tool is used mainly for documenting the various processes involved in a physical system.

2.6 FUNCTIONAL FLOW DIAGRAMS

A type of flowchart which is useful for representing the flow of documents through the various functional segments of an organization is the functional flow diagram. The functional flow diagram in Figure 2–15 shows those segments of the organization responsible for the various activities, and the flow of documents from one segment to another. It uses the same symbols as other flowcharts and is very effective when analyzing manual systems. Again the flowchart is used as a tool to describe physical systems — in this case a manual system.

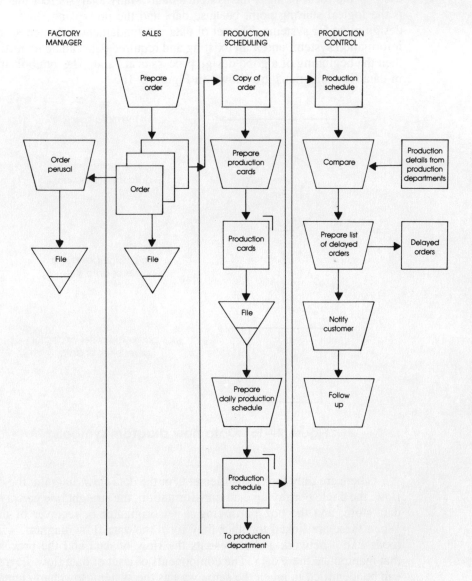

Figure 2—15 Functional flow diagram

2.7   DATA FLOW DIAGRAMS

Data flow diagrams (DFD) are a more recently developed tool generally accepted by the system development community. This, no doubt, is due to their simplicity and inherent flexibility for the representation of data flow in a system. No attempt is made to show the equipment or media used — the focal point is the system's data. Many analysts feel that this is the logical starting point because data and the processing flow is the design pivot for systems. The set of data is a fundamental element of any information system, and if all existing and required data can be identified, then the beginning of a good design process is at hand. The symbols used in data flow diagrams are shown in Figure 2–16.

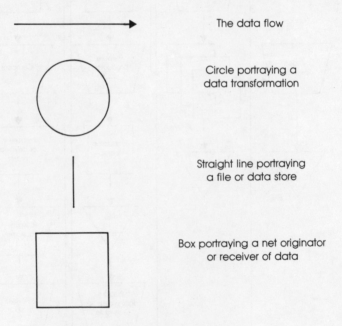

Figure 2—16   Data flow diagram symbols

There are only four basic elements in the data flow diagram: the data flow, the circle portraying data transformation, the straight line portraying data store, and the box portraying a net originator or receiver of data. When these are linked together they form the data flow diagram, which looks like a network and represents the flow of data and the processes that manipulate these data. The components of a set of data flow diagrams can be identified in much the same way as the system flowchart elements were in Figures 2–9 to 2–14. Usually a system can be represented in its simplest form by an overview diagram which shows the major inputs and outputs, for example in Figure 2–17 the major inputs and outputs to a fixed asset register system are shown.

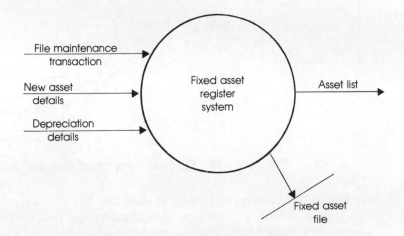

Figure 2—17   Overview diagram

Figure 2–17 shows how the major inputs and outputs of a system can be represented in an overview diagram. It should show the minimum data consistent with this objective. At the next level (level 0), the major parts of the system are shown. For example the fixed asset register system might consist of the processes:

- Calculate depreciation charge
- Process new asset details
- Process file maintenance transactions
- Produce report

The data flow diagrams to represent these elements would show the origins of the data, the data flows, processes and receivers of the data. An example of part of the level 0 diagram is shown in Figure 2–18. This diagram identifies one of the major processes in the system and shows the relevant data flows.

Figure 2–19 shows the complete level 0 logical data flow diagram for a system which processes function bookings and billing for a club. In this system customers make either an inquiry, a booking, provide firm details of bookings or cancel a booking. These data are labelled function transactions on the data flow diagram. A clerk then directs the data ac-

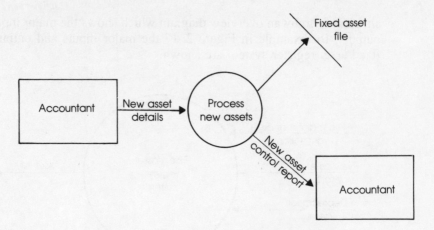

Figure 2—18   Process new asset details

cording to their nature, after making sure that the data are suitable for this system. Inquiries, bookings, cancellations and firm details are all entered in a master diary which contains details of facilities available in chronological sequence. Using this master diary, booking confirmations are possible as well as the eventual billing of the customer. Before billing can take place, however, details of resources used during the function are entered into the master diary from resource details provided by the catering department. A further part of the system is the preparation of weekly activity details for the catering department so that they can plan the functions coming up in the ensuing week.

This is called a logical data flow diagram because none of the physical details of the system have been shown in the diagram. Physical details can be shown on a data flow diagram by including information such as the names of people currently involved in processes in the system, or details such as the sequence of the master diary. From a logical point of view however, this information adds nothing to the description of the system.

Taking each of the four symbols used in the diagram the following points should be noted. Each data flow carries a unique name except for simple flows to or from data stores. Thus the data flow labelled 'inquiry' has a different composition to the data flow labelled 'inquiry data'. If the composition were exactly the same then the same title would be used on the diagram. This implies that when drawing a data flow diagram you need to know the exact composition of each of the data flows. These details are normally carried in the data dictionary which is discussed in Chapter 4. The names selected for the data flows should also be inform-ative — allowing the data flow diagram to act as a communication vehicle. It should be noted that the data flow line only represents data; no flow

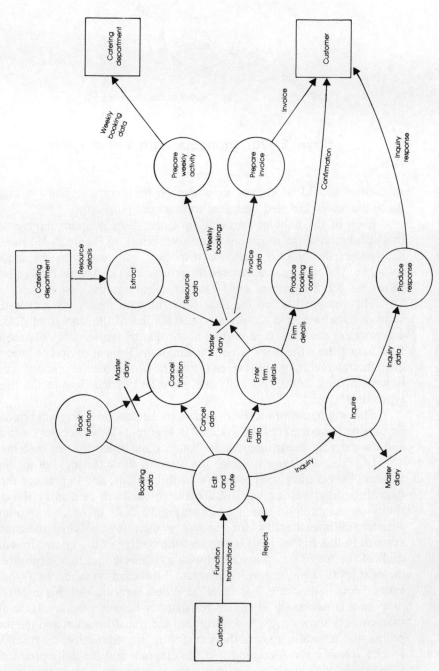

Figure 2—19   Sample logical data flow diagram

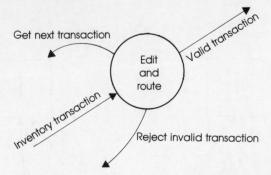

Figure 2—20   Erroneous data flow diagram

of control should be shown anywhere on the diagram. Thus in Figure 2–20 the arrow labelled 'get next transaction' is in error.

Each of the bubbles (circles) on a data flow diagram represents a transformation of the incoming data flows. Thus, in Figure 2–19, inquiry data enters the process 'produce response' and results in the inquiry response. The raw data has entered the process and a response (in a suitable form) has left the process and in this case gone to the customer.

The straight line on the data flow diagram represents a file, data store or data base; it is simply a representation of the storage of data on a permanent medium. It should be noted that by convention only the net flow to or from a file is shown on the diagram. Thus in an update process, even though records would be read from the file before the updated record is written back to the file, only the net effect or the flow to the file is typically shown.

The originators or receivers of data in the data flow diagram typically lie outside the context of the system. In Figure 2–19 we see the customer and the catering department as the sources and sinks for this system.

In Figure 2–19 one diagram was used to show the logic of a simple system. This is merely an overview of the system, and in practice many data flow diagrams are used in a hierarchical form to display the data flows. An example of this is given in Figure 2–21 in which a two-level structure outlines the flows for a simple production scheduling and control system. In this figure, level 0 provides the outline of the system in which each of the major functions are shown as a circle and then exploded in the next level. Thus, at level 1, diagram 3 (production scheduling) shows further details of circle 3 at level 0. When carrying out this explosion process it is necessary to ensure consistency between the levels. In this case circle 3 shows copy A coming into the transformation and the daily production schedule leaving the transformation. Similarly, diagram 3 at level 1 shows copy A coming into the diagram and the daily production schedule leaving the diagram. Note the numbering system that is used at the different levels. If, for example, further explosion of circle 3.2 was

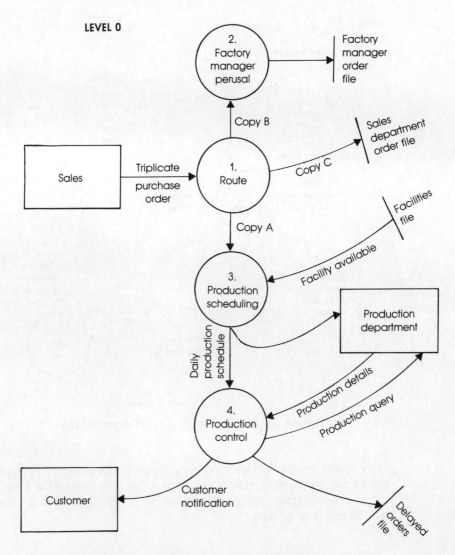

LEVEL 0

Figure 2—21 (a)   Data flow diagram: level 0

considered necessary at level 2, then the circles would be numbered 3.2.1 and so on.

To summarize, the data flow tools graphically show data flows without placing emphasis on the physical or control aspects of the system. The diagram does not reveal any timing considerations in that system. These attributes provide some significant advantages to the systems analyst/designer because they allow the logical system attributes to be considered quite separately from the physical attributes. This is partic-

LEVEL 1

3. PRODUCTION SCHEDULING

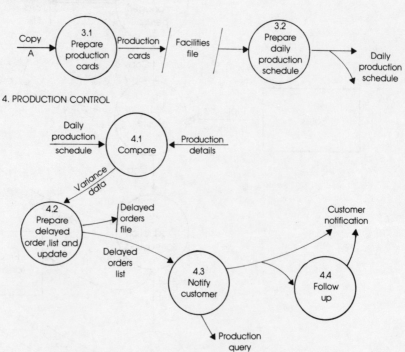

4. PRODUCTION CONTROL

Figure 2—21 (b)   Data flow diagram: level 1

ularly useful during the development process as the designer attempts to understand an existing system or to plot the behaviour of a proposed system. The diagrams are graphic, partitioned, simple, and they concentrate on the flow of data.

2.8   HIERARCHY PLUS INPUT-PROCESS-OUTPUT (HIPO)

Hierarchy plus input-process-output (HIPO) was developed by IBM as a design aid and documentation technique which attempts to:

1. Provide a structure by which the function of a system can be understood
2. State the functions to be accomplished
3. Provide a visual description of the input, process and output for each function

A HIPO package consists of a hierarchy diagram and input-process-output diagrams. An example is given to clarify the approach.

Consider the simplified computer-based inventory system illustrated by the system flowchart in Figure 2–22. Figure 2–23 shows the hierarchy diagram for this system. This diagram acts as a hierarchy chart for the function to be performed and contains a reference, on the boxes, to the number of the diagram which displays the detailed processing descriptions in terms of an input-process-output diagram. Figure 2–24 presents the overview (IPO) diagram in which the inputs, major processing functions and outputs are displayed. As an example of a lower level IPO diagram, Figure 2–25 illustrates the edit stock issues process.

This example indicates that HIPO is hierarchical, like data flow diagrams and can be used to represent either a logical or physical view of a system.

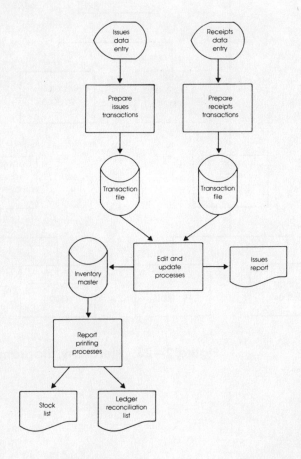

Figure 2–22   System flowchart

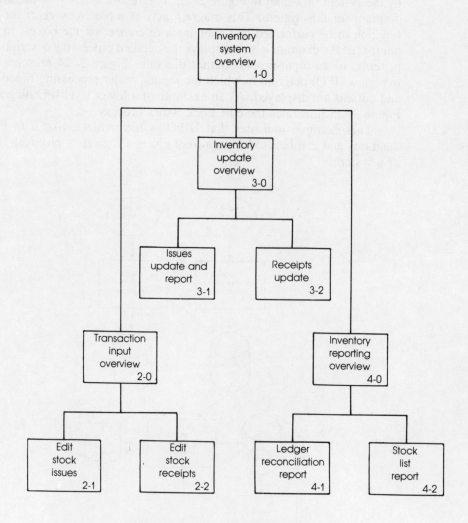

Figure 2—23   Hierarchy diagram

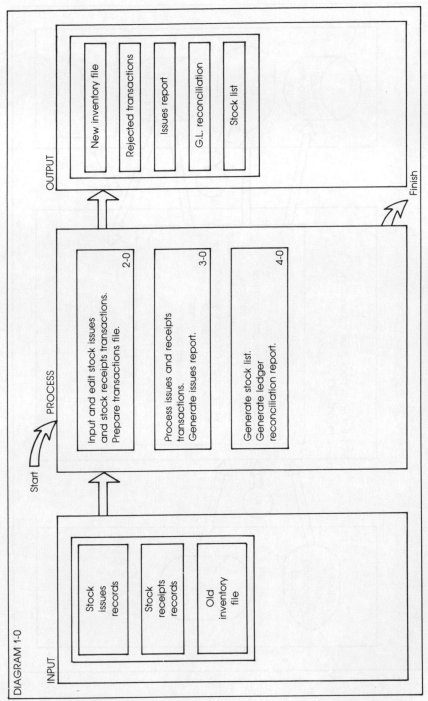

Figure 2–24 Overview diagram

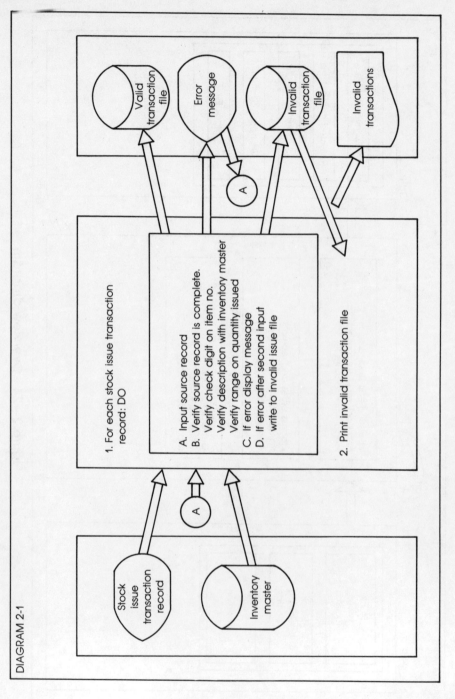

DIAGRAM 2-1

1. For each stock issue transaction record: DO

   A. Input source record
   B. Verify source record is complete.
      Verify check digit on item no.
      Verify description with inventory master
      Verify range on quantity issued
   C. If error display message
   D. If error after second input
      write to invalid issue file

2. Print invalid transaction file

Valid transaction file

Error message

Invalid transaction file

Invalid transactions

Stock issue transaction record

Inventory master

Figure 2—25   Example detail

2.9   REVIEW QUESTIONS

1. Describe the relationship between system and logic flowcharts.

2. How is a data flow similar to a flowchart? What major differences exist?

3. (a) Convert the functional flow diagram Figure 2–15 into a data flow diagram.

   (b) What additional data do you think would be necessary to allow the processes shown in your diagram to be completed in practice?

4. Figure 2–20 shows an erroneous data flow diagram. Devise another diagram which contradicts some other rule of data flow diagrams.

5. (a) Expand the circle 4.2 in Figure 2–21(b) to show a level 2 representation of the process. Assume:

   (i) Only key customers are notified. The customer file contains a key customer indicator

   (ii) Only delays in excess of three days are notified

   (b) Should the customer file be shown in the level 1 diagram as input to circle 4.2?

6. Draw a data flow diagram to represent the information in Figure 2–22.

7. Why is a double-headed arrow used to connect the VDU symbol to the process symbol in Figure 2–10?

8. What activities are performed in the following system functions?

   (a) Data entry
   (b) Edit
   (c) Update
   (d) Report
   (e) Rearrangement

9. What information can be more readily displayed using HIPO diagrams rather than data flow diagrams?

10. **Data Flow Diagram Case Study**

    The purchasing sub-system described below does not involve any computerization. At present the following procedures apply:

    • At the beginning of each year vendors for standard items are reviewed on the basis of price, past service and payment terms, and a list of approved vendors is prepared by Bruce Hassel, the purchasing officer. This list is then approved by the factory manager.

Purchase requisitions are raised by operating departments and sent to the purchasing section which enters each requisition into a purchase requisition register under purchase requisition number.

When a purchase requisition for a standard item is submitted to the purchasing section by one of the operating segments, one of two procedures is followed. If the cost is $1000 or more the purchase requisition must be approved by the factory manager; if less than $1000 the purchasing officer approves. After approval and selection of a vendor (vendor code entered), all requisitions are sent to the accounting section where the account to be charged is entered as a code and the requisition returned to purchasing. Any requisitions not approved by the factory manager or purchasing officer are returned to the originating operating segment with the reason for disapproval written in.

Purchase requisitions for non-standard items are prepared on a special requisition form and must be approved by the factory manager. If the anticipated cost of the order is less than $1000, then Bruce Hassel selects a vendor and records the vendor's name and address on the requisition. If a requisition is for $1000 or more, Hassel prepares a 'request-for-bid' form which is sent to two or three possible suppliers, and the lowest bidder is accepted and entered on the requisition.

- After the purchase requisition is returned from the accounting section, a prenumbered purchase order is prepared and distributed as follows:

  (i)   Original to the supplier, after updating purchase requisition register

  (ii) Copy to open order file with purchase requisition (filed alphabetically by vendor) and entered in the purchase order journal (held in the purchasing department) in order number sequence

  (iii) Copy to originating department

  (iv) Copy to receiving store.

- When the goods are received, a receiving report is prepared by the receiving store showing the supplier's name, items received, and quantity — indicating any damage or deficiencies. The original of this goes to purchasing and a copy is filed (alphabetically by vendor) by the receiving store.

- In the purchasing department the receiving report, purchase order and invoice are compared for quantity and price, and if satisfactory are sent to the accounting section. If a price or quantity difference exists, the purchasing department initiates an investigation into the reasons for the difference. When orders are completed, an entry is made in the purchase order journal to indicate the completion. Each month the purchasing department prepares reports on:

  (i)   Outstanding requisitions listing all requirements for which an order has not been placed

(ii) Outstanding orders listing all purchase orders unfilled

(iii) Orders received report showing all orders fulfilled during the month.

Questions

(a) Express this logic in a set of data flow diagrams.

(b) Express this logic using a functional flow diagram.

(c) Discuss any difficulties you perceive with this system.

3 Procedure Representation Tools

3.1 INTRODUCTION

This chapter looks at some of the tools available to the system designer to represent the logic of a particular procedure or process. This representation may assist in the implementation stage when the process is typically converted into a program or set of programs. A process is a definable set of logic used to convert a set of input into a set of output. An analyst will need to understand and document processes in many stages of the system's lifecycle. For example, to analyze the current system it is generally necessary to document it using a selection of the tools presented to throw light on the activities of the organization. The techniques described in this chapter are:

- Flowcharts
- Decision tables
- Pseudocode
- Structure charts
- Flowblocks

These five tools provide a representation of those commonly used from the 1960s to the present. However, it should be borne in mind that in many cases the logic of a procedure may be so trivial or so well known that it is not necessary to document it. It is also not uncommon for the requirements of a process to be expressed in natural language. Quite often a representation of the input and output flows of the process, along with an English text description of its objectives and logic, are sufficient to specify the procedure. Where this is possible it has the obvious advantages of minimizing the effort in specifying the task, and making the documentation more readable to the user. While English text may be suited to documenting relatively straightforward logic it can become ambiguous and inefficient for specifying a complex process.

3.2 FLOWCHARTS

Logic flowcharts were introduced in Chapter 2 as a means of describing one of the many interconnected operations in a given information system. Thus the logic flowchart is a micro-expression dealing with only a portion of the overall system. Essentially a logic flowchart consists of a network constructed from outlines, which represent certain operations, interconnected by directional lines indicating the next operation in the sequence. In a logic flowchart one box is designated as the first to be processed, the entry point; another as the last, the exit point. These points are usually described by a rounded box, as shown in Figure 3–1. Other boxes describe decisions, processes or connectors to other flowcharts. These symbols are also shown in Figure 3–1.

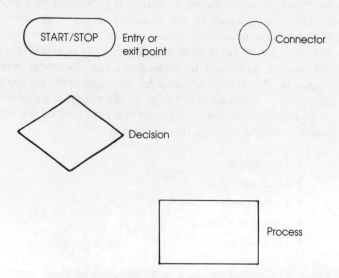

Figure 3–1 Flowchart symbols

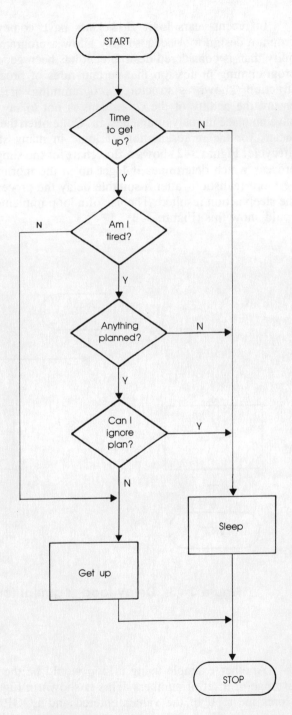

Figure 3—2   Logic flowchart

In recent years logic flowcharts have come under criticism as a program design tool because they allow a program designer more flexibility than is considered desirable. It has been generally accepted by the programming profession that certain rules of program design and construction (known as structured programming) are desirable in order to ensure the quality of the code. This is not to say that logic flowcharts have no place in analysis and design. Quite often they provide an excellent means for the representation of logic in many stages of the system's lifecycle. Figure 3–2 shows a flowchart for the simple logic of a decision process which determines if I get up in the morning. This logic would be more realistic if after a suitable delay the process could start again if the sleep action resulted. The use of a loop implemented with connectors could show this (Figure 3–3).

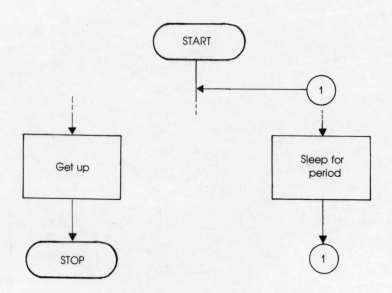

Figure 3—3   Delay loop segment for Figure 3—2

Another example using a loop would be the logic to calculate the average of a set of numbers. This is shown in Figure 3–4 — here SUM stores the total of the values entered and LOOP stores the number of values entered.

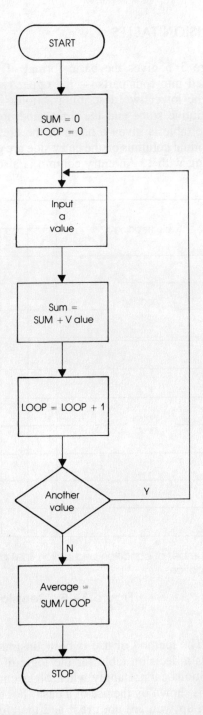

Figure 3—4   Logic flowchart with loop

3.3 DECISION TABLES

Figure 3–5 gives the basic format of the decision table. The table is divided into four parts — the upper two sections describing conditions and the lower two describing actions. The left-hand sections are called descriptive stubs and the right-hand are called entries. In addition, the whole table is given a name in the area labelled table heading, and the individual columns on the entry side are given identifiers, usually numbers starting with 1. An entry column is also called a rule.

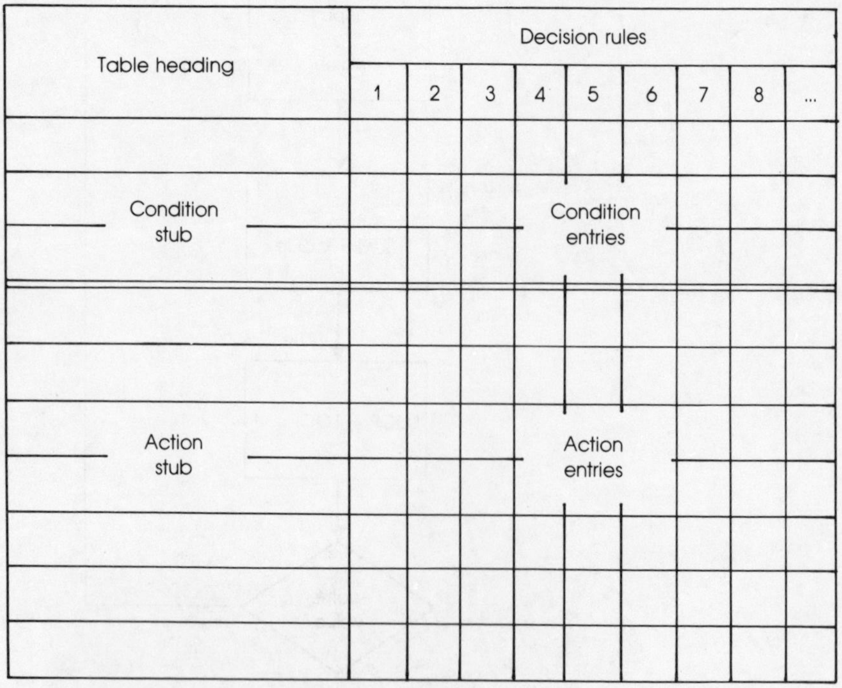

Table heading	Decision rules								
	1	2	3	4	5	6	7	8	...
Condition stub				Condition entries					
Action stub				Action entries					

The number of condition and action lines as well as rules is adjusted to suit the application

Figure 3—5 Decision table format

The method of use is best illustrated with an example. Figure 3–6 shows a decision table for the logic of Figure 3–2. Notice how certain conditions cannot apply when other conditions are in a particular state. This is shown by the use of a dash. For example, in column two it is time to get up, you are not tired, and therefore get up regardless of the other conditions.

Get up decision table	1	2	3	4	5
Time to get up?	Y	Y	Y	Y	N
Am I tired?	Y	N	Y	Y	–
Anything planned?	Y	–	N	Y	–
Can I ignore plans?	Y	–	–	N	–
Sleep	X		X		X
Get up		X		X	

Figure 3—6   Decision table task example

3.4   PSEUDOCODE

Pseudocode is a non-graphical, loosely defined means for expressing program logic. It bears a strong resemblance to many high level programming languages, featuring block indentation and formal English style statements. The key words used reflect the preferred control constructs such as if, while and until. The major difference between pseudocode and a formal programming language is the use of less formal explicit English expressions instead of the formal syntax of a programming language.

An example of pseudocode is:

```
IF light is red
    Stop
ELSE Go
ENDIF
```

Note the use of indentation to specify that the action 'Stop' is under control of the IF, and also note that the ELSE and ENDIF are aligned with the IF to signify that they are part of that construct. Another example might be:

```
PERFORM UNTIL Stop
    Read a record
    IF end of file
        Stop
    ELSE IF Balance on hand < reorder point
        Print reorder
```

Do you recognize this logic from Chapter 2?

Many different forms of pseudocode have been developed, usually to reflect the many programming languages which are available today. Of course there is nothing to stop an organization using its own internally developed pseudocode by adapting a suitable structured language known to all in the organization. Figure 3–7 gives an example of pseudocode.

PERFORM UNTIL I get up

    SLEEP FOR PERIOD

    IF I am tired

        IF anything planned

            IF I cannot ignore plan
            get up

            ENDIF

        ENDIF

    ELSE get up

    ENDIF

END PERFORM

Figure 3—7   Pseudocode for get up

In this logic the process is controlled by an UNTIL which basically says until you get up, sleep for the delay period. After the sleep you carry out three tests. If you are not tired, or if you have a plan and cannot ignore it, you get up. Otherwise you are still under control of the PERFORM UNTIL and therefore you sleep again.

3.5 STRUCTURE CHARTS

Some program design theories hold that the structure of the program should reflect the structure of the input, output and files that the program is handling. This section looks at the means of representing the structure used in these theories. Three basic structure types are used as shown in Figure 3–8; these are the sequence, the iteration and the selection structures.

Each of the structure types is made up of components which are described as elementary components. An elementary component is one which is not dissected to any further level in the diagram. A sequence structure has a number of parts, each of which is executed or appears once only and in order. For example, a program may consist of a sequence of three paragraphs of code B, C and D, which are executed once in that order. Alternatively, a file of data might consist of a sequence of one B-type record, followed by one C-type record, followed by one D-type record. An iteration structure is performed or occurs zero or more times. For example, A might consist of a paragraph of code B which is performed a number of times.Typical programming constructs used to execute an iteration are DO, DO WHILE and PERFORM . . . times. Examples of iteration data structures are a file consisting of one or more records, or a variable length record having zero, one or more fields of a certain type. A selection structure has a number of parts, only one of which is present in each occurrence of the structure. The case construct and IF condition are typical examples of selection in a program. In data, a selection component may be used to represent a valid or invalid record, or a record might be a credit note or an invoice in a transaction file.

An illustration of a program structure is given in Figure 3–9. The program is to print an edit report consisting of three parts in sequence: the report heading, the report body and the total line at the end of the report. The report body consists of an iteration of lists for each batch which has been edited and each batch consists of a batch heading, an iteration of record lines (each of which is either valid or invalid) and a batch total.

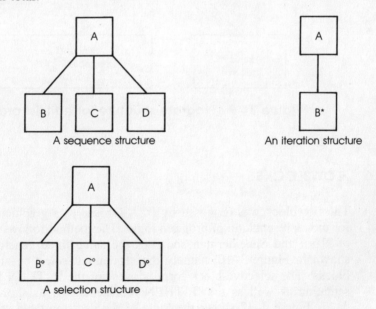

A sequence structure

An iteration structure

A selection structure

Figure 3—8   Basic structure types

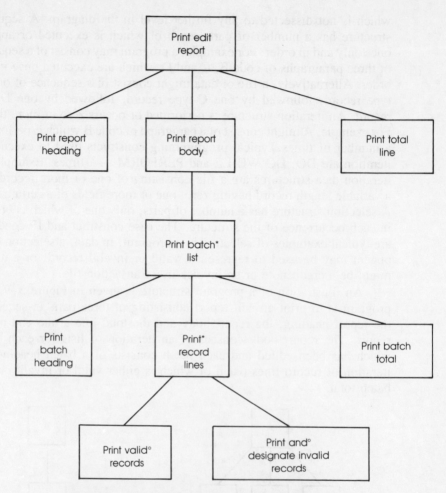

Figure 3—9   Program structure for edit list program

3.6 FLOWBLOCKS

The flowblock was proposed by P. J. Grouse as a graphical alternative for the representation of program logic. The method follows on the work of Nassi and Shneiderman and uses the same three general constructs shown in Figure 3–10: namely the sequential, selection and repetition blocks. The selective block form is used for the IF THEN ELSE representation as well as the IF THEN representation, as shown in Figure 3–11. Figure 3–12 shows the logic of the decision table of Figure 3–6 put into a flowblock form. We see here that the true block is the one on the left and the false block the one on the right. Note that the flowblock includes the logic for the delay loop, and introduces the UNTIL construct to facilitate the STOP and the loop.

This form of representation can be used for high level specification of logic or at a very low level for something akin to pseudocode. You will notice that there are certain similarities between flowblocks and pseudocode. The major difference is that the flowblock uses both vertical and horizontal dimensions to convey information (such as in the side-by-side placement of alternatives in selective blocks) whereas the pseudocode uses only vertical placement. Consequently some argue that the flowblock is more easily understood.

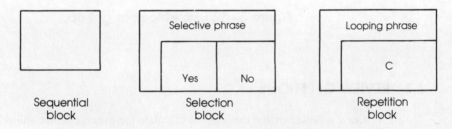

Figure 3—10   The three general flowblock constructs

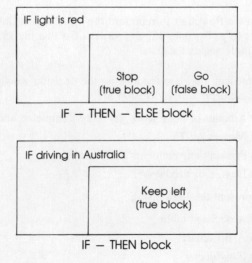

Figure 3—11   Selection constructs

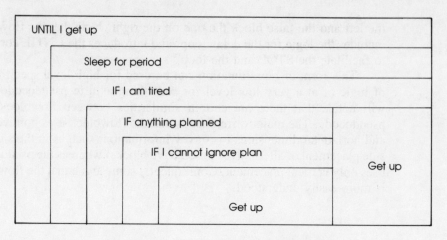

Figure 3—12   Flowblock of get up

3.7   REVIEW QUESTIONS

1. Draw a flowchart for the logic to calculate the average of six values.

2. Modify the flowchart drawn in question 1 to handle any number of values.

3. Modify the flowchart drawn in question 1 to calculate the standard deviation of the six values.

4. Draw a flowchart to represent the logic of merging two sequential numeric lists (e.g. list one: 17, 34, 48, ...; list two: 8, 15, 36, 68, ...) to produce a single sequential list.

5. Represent the logic for making the decision whether to go to bed. Assume the elements in the decision are:

   (a) Amount of homework yet to be completed and its importance

   (b) Movies on TV

   (c) Pressing engagements early next morning

   (d) Degree of tiredness

   Represent the logic using:

   (a) A decision table

   (b) A flowchart

   (c) Flowblocks

   (d) Pseudocode

6. A capital asset file holds the following information for each asset:

   • Asset number
   • Original cost
   • Estimated life
   • Yearly depreciation percentage

   Write the pseudocode to show the instruction sequence for calculating the depreciation for each item, and total depreciation for all items on the file.

7. You have just been appointed as a tutor in information systems and each week you find it is necessary to calculate the mean assignment mark for the students in each of your tutorial groups. As the number of groups you take varies, as does the number of students in each group, you decide you will write a program to calculate this average. This interactive program should allow you to enter from a visual display unit or terminal (VDU) any number of marks and calculate the average when desired.

   Draw a logic flowchart for this program.

8. A used-car firm maintains a file of cards, one card for each car sold in the past year. The manager wishes to have a program he can run to tell him how many cars of a certain year and make were sold by a particular salesman and what was the average gross margin of these sales.

   The file contains:
   • Car make
   • Year of manufacture
   • Salesman's name
   • Cost price
   • Selling price

   The manager wishes to know how many 1980 model Toyotas were sold by Mr McClelland (as well as the average margin). The manager is not familiar with computer terminology or any computer language. Prepare a flowchart which could be used to explain the logic of the required program to him.

4 Data and the Data Dictionary

4.1 INTRODUCTION

Since information systems are concerned with manipulating data as it moves through the various processing stages of input, storage, retrieval and output, the precise definition of the data in these stages is an important step in systems analysis. The repository of these definitions of data is called the data dictionary, and it forms an essential partner with the data flow diagram in documenting and designing systems.

Data is conceived as being composed of data elements which are grouped into data structures. A data element is a non-decomposable unit of data for which values are collected or held. The following are examples of data elements:

- Customer name
- Invoice number
- Price
- Stock on hand
- Book title
- Year-to-date sales value

The term non-decomposable is used to indicate that a data element is not decomposable into lower level data elements in the system being described. For example 'book' is not a data element if it is described as being composed of the data elements:

- Standard book number
- Title
- Author
- Call number

'Book' is an example of a data structure, which is defined as a collection or grouping of logically related data elements. Further examples of data structures can be seen in data flows and files. A data flow is a data structure which passes along a particular arc of a data flow diagram while a file contains a collection of data structures.

The data dictionary can serve to document the data descriptions in both the logical and the physical design stages. In the logical design stage, physical details such as length of fields, coding schemes, physical names and physical representations of the data are omitted. In the physical design stage these are added. To enable information to be easily added to entries and to facilitate keeping entries in alphabetical sequence, a card file is recommended for a manual data dictionary. This would be set up to contain one data dictionary entry (i.e. a data element or data structure) per card. For other than a small system, a computer-based editor or data dictionary presents considerable advantages.

4.2 DOCUMENTING DATA ELEMENTS AND STRUCTURES

The conventions to be used in this book in documenting and describing data in the data dictionary are given in this section. Other conventions are possible. See De Marco (1980) for an alternative representation.

Those proposed in this section are to be interpreted more as suggestions than rules: individual situations require modification. However, when deciding to use a different convention, consideration should be given to those who will have to read and work with the document. Thus many changes from the existing convention should be avoided unless compelling reasons exist for their adoption.

Data elements are identified by a descriptive name of one or more words, and if included in the data dictionary will be accompanied by a brief description. (Hyphens may be used in multi-word names but are not mandatory.)

Invoice value
> Grand total value of invoice after product discounts have been deducted. Query values over $6000.

The above example shows two types of description for the data element 'invoice value': a general description of the element and an editing comment.

Coding structures may also be described.

Customer number
> The existing code is a two-digit alpha (the first two digits of the customer's name) followed by four numeric digits.

If the description of the data element is self evident (e.g. 'company name') then the data element need not be entered in the data dictionary.

For discrete-valued data elements (e.g. 'customer classification' or 'product type') the number, or approximate number, of discrete values can be indicated as a guide to later code design. For example the data element 'product type' may be documented in a data dictionary with the entry:

Product type
> At most eight codes are required.

A data structure is a named collection of logically related data elements. For example 'address' can be viewed as a data structure consisting of street, suburb, state and postcode. The convention used in this book for displaying a data structure is illustrated below:

Address:
> Street
> Suburb
> State
> Postcode

The colon and indentation is used to indicate that 'address' is the data structure consisting of the indented elements. Should 'address' have been described as a data element, one could not refer to a part of it such as 'postcode'.

Data structures can be nested within data structures to any depth. For example the data structure 'cash receipt' may consist of two data elements and two data structures as shown below:

Cash receipt:
> Date:
>> Day and month
>> Year
> Amount
> Customer name
> Customer address:
>> Street and suburb
>> Postcode

Repeating fields

A data element or structure which repeats two or more times may be indicated by means of an asterisk after the name. For instance an invoice may have a number of invoice lines on it. This data structure could be documented as:

> *Invoice:*
> Invoice number
> Date
> Customer name
> Customer address
> Invoice line:*
> product number
> product description
> quantity
> unit price
> extended value
> Total value

Consider the example of documenting the data structure of a recipe. A typical Australian recipe might appear thus:

> *Pavlova:*
> Egg whites   8
> Sugar   2 tbsp
> Vanilla   1/4 tsp

Beat egg whites until stiff. Gradually add sugar while continuing to beat. Then add vanilla. Cook for 50 minutes in a 150°C oven.

It can be seen that the recipe consists of a recipe name, a number of ingredients and quantities, and a preparation description. To document the data structure a name needs to be given to the two elements, ingredient and quantity, and this marked with an asterisk to indicate a repeating field. This structure will be called 'recipe line'. Thus the data structure of recipe is:

> *Recipe:*
> Recipe name
> Recipe line:*
> ingredient
> quantity
> Preparation

An alternative method to document the recipe data structure to avoid giving a name to the two elements, ingredient and quantity, is to enclose them both in parentheses and mark them with a single asterisk as shown below:

> *Recipe:*
> Recipe name
> (ingredient
> quantity)*
> Preparation

This is a useful approach when the creation of a new data structure is not desired.

Note that if the data structure had been documented as in the following example it would imply a different structure than that of the recipe book:

> *Recipe:*
> Recipe name
> Ingredient*
> Quantity*
> Preparation

This structure implies that a recipe has a number of ingredients and an independent number of quantity fields. Thus one ingredient and five quantities would be a possible instance or realization of this data structure. Clearly each ingredient must have only one quantity and thus this data structure is incorrect for the recipe book example.

Optional data

Optional data may be indicated by means of square brackets. In the example below 'ship to address' is optional.

> *Invoice:*
> Invoice number
> Customer name
> Address
> [Ship to address]
> Invoice line:*
> Total value

It is assumed in the above example that the data structure 'invoice line' has already been defined elsewhere in the data dictionary with an entry such as:

> *Invoice line:*
> Product code
> Quantity
> Unit price
> Extended price

Thus it may be referred to by name and need not be defined again. The use of the colon serves to designate a data structure and indicate that if not accompanied by a description, one appears elsewhere in the data dictionary.

Alternative data

Two or more data elements or structures may be alternatives with only one to be chosen in any given instance. For example either a customer number or customer name may be used to uniquely define the customer on a payment receipt. This can be indicated by enclosing the alternate names using the less-than (<) and greater-than (>) symbols, as shown:

< Customer number
Customer name >

However if this approach is not feasible, or if additional explanation is needed, a note in the data dictionary can be used. For example if the alternative to customer number is customer name and address:

Customer number ①	① either customer number
Customer name ①	or customer name and
Customer address ①	address

Extensive notes may at times be required to explain the conditions under which options are selected. Consider the case of the data structure for an enrolling accountancy student. Assume the following conditions apply:

- If the student is enrolling for the first time he will not have a student number
- If the student is re-enrolling he will indicate student number and name and address
- Full-time students in first or second year take four subjects each semester, while a full-time student in third year takes three subjects each semester (the conditions relating to part-time students have been ignored for simplicity)

This data structure may be represented as follows:

Student enrolment:
Student number [1]
New or re-enrolment
Name
Term address
[Home address]
Full or part-time
Year or stage
Subjects enrolled semester 1[2]
Subjects enrolled semester 2[2]

Note: [1]If new enrolment no student number is supplied.
[2]The number of subjects to be taken in each semester is given in the table below.

Year	Full- or part-time	Number of subjects each semester
1	Full-time	4
2	Full-time	4
3	Full-time	3

Data naming conventions

The data flow diagrams and the data dictionary are much easier to read and follow if data elements and data structures are called by their common names. However, the common names may not be unique and some additional identification may need to be attached to avoid confusion.

For example, 'cost' may be defined separately in three different departments for which one integrated system is being developed. The names chosen may be cost 1, cost 2 and cost 3 or something like cost-direct, cost-factory and cost-full.

Another situation arises when the same data element or structure is called different names in different departments. For example 'cost 1' in the factory may be 'direct cost' in the accounting department. Rather than force one department to change what may be a well-established name, an alias definition can be set up in the data dictionary. For the above example two entries would be made for the data element, as shown below:

Cost 1 — alias direct cost
    Equal to the direct cost of materials and labour without factory or general overheads
Direct-cost — see alias cost 1

4.3   FILES

Since a file is a static collection of stored data structures it can be represented in the data dictionary using the techniques already described. For example a student file may be documented as:

> *Student file:*
> Student name
> Student address
> Course
> Results:*
> year
> subject
> grade

If the data on the file has already been described separately as the data structure 'student history' then the entry for student file could be written as:

> *Student file:*
> Student history:

Files are iterations of records, though it would not be standard practice to use the iteration symbol on the student history data structure above. It would be redundant information as it is implied by naming the entry a file. Data dictionary entries for files are usually grouped together into a single section.

Two types of files are described in a data dictionary: transaction files and master files. Transaction files are used for temporary storage of data structures. For example, invoice transactions may be stored on a serial file for later processing by an accounts receivable system.

Transaction files usually contain one record per transaction with the records written and read serially on the file. Serial denotes records which are sequential though they do not have record keys in sequence. However, since the data flow diagram and data dictionary are logical concepts we do not need to confine transaction files to be processed serially. They can also, at a logical level, be directly accessed as in the example below where students failing a particular course are extracted (it is not of concern how) and a report written called 'failed student list'.

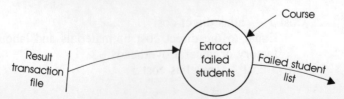

Obviously where specific records must be extracted, the selection conditions must be communicated to the process. In the example above 'course' acts as the logical key to access the transaction file and extract the data structures of those students undertaking the named course. The process 'extract failed students' examines the grade of each of these students and reports those who have failed.

Master files differ from transaction files in that they hold permanent data which is generally updated to reflect changes. Transaction files are by contrast not updated. To update a master file sufficient data must be present with the updating transaction to uniquely locate the right record. For example the data flow diagram and the accompanying data dictionary shown below cannot function as not enough data is present in the transaction to identify the correct customer record. The data flow 'payment' does not contain any customer reference, such as customer number or customer name, to act as a logical key to access the file. It is not sufficient that customer number was known in process 1; it must be available to process 2 which controls the file access to customer file.

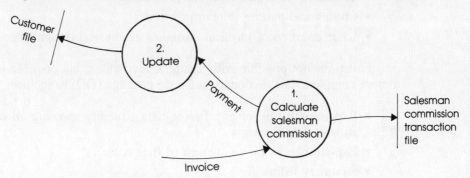

Customer file:
    Customer number
    Customer name and address
    Balance owing
Invoice:
    Customer number
    Customer name
    Invoice number
    Payment amount
    Salesman number
Payment:
    Payment amount
    Invoice number

If 'payment' is redefined to also include 'customer number' the data flow diagram can function.

4.4 AUTOMATED DATA DICTIONARY

Building, maintaining and interrogating a data dictionary can be a very time-consuming task if done manually, and so is a prime candidate for automation. A further significant reason for automating the data dictionary in a large project is the need for it to be a readily available central reference for all project team members during development. It is difficult to achieve this in a manual system.

A simple means of data dictionary automation in a small to medium-sized project, when a specialized data dictionary system is not available, is to use a text editor. This would enable the following functions to be easily carried out:

- Alphabetical insertion of data elements or data structures
- Addition to or modification of entries to allow top-down development of the data dictionary
- Searching for a particular entry by name
- Cutting and pasting to rearrange order
- Later insertion of physical attributes at physical design stage

Functions not possible with a simple text editor, but possible with a more complete data dictionary software package (DDSP) include:

- Expanded listing of any file or data structure showing all constituent data elements
- Explicit checks on uniqueness of data name
- Summary listing
- Extract listing (e.g. list the contents of the data flows in the personnel sub-system)
- Where-used cross reference listing
- Consistency and completeness checking (e.g. verify that data elements output from a process can be derived from data inputs)
- Generation of high level language data declaration statements

A number of the functions listed above require the DDSP to have access to not only the data dictionary, but also the data flow diagrams and the process descriptions.

This can be seen by considering the task of consistency checking. In order to verify that output data elements from a bubble can be derived from input data elements, the logic of the process and the identity of the input and output data elements must be known. A number of DDSPs are available commercially that accomplish most of the functions listed above.

4.5   REVIEW QUESTIONS

1.   What is the relationship between a data dictionary and a data flow diagram?

2.   List and define the data dictionary conventions described in this chapter.

3.   Define a data dictionary entry for the works time sheet shown below.

Excelsior Manufacturing Co. Pty Ltd
EMPLOYEE'S TIME SHEET

Date...

Clock No..Name..

Time		Description of job	Eng. trade *	Job no.	Ord. hrs.	O/time hrs.	Remarks
Start	Comp.						

*This is only supplied on maintenance jobs

4.   Define a data dictionary entry for the purchase requisition shown below.

		Rossichal Manufacturing Pty Ltd PURCHASE REQUISITION		

PR No. 1732

Date..

Supplier..Ship to..

....

Date required..

Department..Charge to...

Quantity	Unit of measure	Description	Cost

Special instructions

Person ordering... Approved...

5.   Design a goods received note for a smallgoods wholesaler, and define its data dictionary entry.

6.   Design a statement form for a retailer of microcomputers and associated software, and define its data dictionary entry.

5 Files

5.1 INTRODUCTION

Chapter 4 looked at the way in which the data of a system might be documented and described in the data dictionary. This chapter looks at the way in which the data is physically stored on files so that it can be used as part of the information system. Files are organized collections of data which, in most systems, are stored on magnetic disk or magnetic tape. One of the most significant factors affecting a system's effectiveness and efficiency is the manner in which data are organized in files and, consequently, this decision is one of the most important in the design of computer-based systems.

The term 'file' is used in a computer context in exactly the same sense as in its general use: for example, a file of correspondence, a file of notes or a file of receipts. Files are necessary for any system, be it manual or automated, in order to provide storage data with the objective of retrieving such data when it is required. Examples of common files in business systems would be an accounts receivable file containing data concerning the debtors of an organization; an inventory file containing data concerning the stocks held by an organization; a product file containing the data relating to products; or an order file containing orders received for goods from the company. Each file is made up of records and each record is made up of data elements. These relationships are shown in Figure 5–1.

A data item or data element describes some particular attribute of data, e.g. a debtor's name, an invoice amount, an employee's age or an employee's net salary. The space reserved for a data item in a record is sometimes termed a field. A record contains all of the data items which relate to a specific object of data processing. For example, a debtor's record in an accounts receivable file might contain the following data items:

- A debtor's unique number
- The debtor's name and address
- The debtor's balance owing
- Credit rating

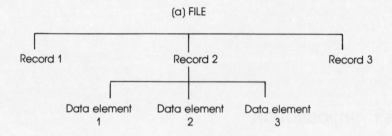

(a) FILE

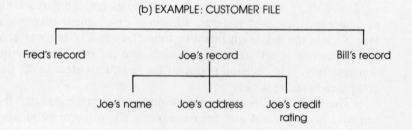

(b) EXAMPLE: CUSTOMER FILE

Figure 5—1   A file structure and an example

Another example is a credit sales transaction record containing the following data items:

- Customer number
- Inventory number
- Quantity sold

5.2   TYPES OF FILES

In order to understand the different types of data which may be stored in computer files, it is usual to sub-divide these files into several categories such as:

- Master files
- Transaction files
- Report files
- Work files

A master file contains records of relatively permanent information which typically describes the status of a particular class of objects. Most organizations have a product master file which describes the products carried or manufactured by the organization, a supplier master file which provides information about each supplier to the organization, a payroll master file which contains details of the payroll, and an accounts receivable master file which gives details of amounts owing to the organization.

Each record in a master file will normally contain at least one data item which acts as the key to identification. Examples of keys are a customer number in the debtor's file, an employee number in a payroll file, a product number in an inventory file, or an order number in an order entry file. The concept of keys is critical to an understanding of computer-based files, as it is the key which provides the basis for both the storage and recovery of records within a file.

Transaction files, as the name suggests, contain data relating to transactions which have occurred against particular applications such as payroll or invoicing. When the data processing activity takes place the master files are brought up to date (updated) by using the data stored on the transaction files. For example, an employee master file may be updated to reflect an employee's change of address, or the stock-on-hand field in the stock file updated after a sale of goods. With transaction files the concept of a record's key does not apply, but attributes of the record (e.g. a customer number in an invoice transaction) are used as the basis for matching the record with the relevant master file record.

Work files are used to provide temporary storage of data whilst processing is being carried out. For example during a large processing operation many temporary files might be created (to hold intermediate data which will be processed further in a later run).

Report files are another category. At times during a processing operation it is desirable to extract certain information from a master file (e.g. the information to be subsequently printed out in a report). It may be more efficient to extract the relevant details from the master file whilst processing is taking place (say, performing an update), write these on to a new report file, and then later print that file in a report format.

5.3   STORAGE DEVICES

In a perfect computer configuration there would be unlimited storage with instant availability. The perfect situation does not exist, however, and the different types of storage available need to be assessed in terms of their suitability for the particular storage task at hand. Table 5–1 shows the types of memory devices available, along with their capacity, speed and cost characteristics.

Device	Acess time	Capacity (bits)	Cost (cents/bit)
Register	2-20ns	10^3 -10^4	20-100
Cache	20-200ns	10^4 -10^5	1-10
Main memory	200-2000ns	10^6 -10^8	0.1-0.5
Secondary storage Head-per-track disk, bubble, drum	1-10ms	10^7 -10^9	0.05-0.2
Removable disk	20-100ms	10^7 -10^{11}	0.0005-0.01
Tape and strip handlers	1-100 sec	10^{11}-10^{12}	0.000 02-0.000 05

Ins = 10^{-9} sec
Ims = 10^{-3} sec

Table 5–1   Memory types

It can be seen from this table that as access time increases, so does the available capacity, but the cost per bit (binary digit) of data stored decreases. Thus it is necessary to accept a compromise between cost, speed and capacity. Another factor affecting the choice is the permanency of the data stored. Only secondary storage and tape and strip handlers categories of Table 5–1 provide a non-volatile medium on which data can be stored permanently without the provision of power. For this reason the term 'auxiliary storage' is used to describe these non-volatile media.

Details of the construction and operation of magnetic tape units, disk drives and other hardware are given in introductory books on computer systems and will not be covered here except in so far as is necessary to explain concepts.

On-line secondary storage (storage which is available to the computer processor) is provided almost exclusively by magnetic disk in today's computer configurations. This is because of the capacity of the disk, the retrieval methods available, the speed of access and the cost of data stored on this medium. Capacity refers to the quantity of data which may be stored on a medium. It is usual to refer to auxiliary storage capacity in megabytes (one million bytes) where one byte is equivalent to 8 bits. Each byte can typically store one character. In some cases capacity is quoted in 'words', each word consisting of a specified number of bits or bytes. As an example of different capacities, magnetic tape comes in various standard densities of 800 bytes per inch (bpi), 1600 bpi or 6250 bpi. Thus, a 2400-foot 1600 bpi tape could theoretically contain over 46 million bytes of data. A certain amount of this capacity, however, is not available for data storage due to the need to provide a space between records or blocks of records.

5.4 FILE ORGANIZATION AND RETRIEVAL

Basically there are two types of file organization: sequential and random. As the name suggests, a sequential file is one in which contiguous records follow a pre-determined sequence. This is usually key sequence, either ascending or descending, but may be in a time sequence or sequenced (say) alphabetically on a field other than the key field. A random file is one in which there is no simple relationship between contiguous records.

A primary objective when deciding the most appropriate method of organizing data in files is to provide write and retrieval efficiency, because access to data on secondary storage is slow compared to other computer functions. The factors which affect the selection of file organization are:

- Volatility
- Activity
- Size

Volatility. This refers to the number of additions, deletions and/or changes to the records in a file.

Activity. The amount of activity on a file is the proportion of records which have to be processed in any processing session. For example, a payroll file has high activity during an update process because practically every record is processed. The activity ratio is the number of records processed compared to the total number of records on the file. This is also referred to as the 'hit rate'.

Size. If a file is very large it is sometimes desirable to alter its organization and processing in order to achieve retrieval efficiencies. For example certain applications use abridged versions of master files at certain times to provide faster data retrieval in on-line real-time systems. The full version of the file might be processed overnight when the system is no longer running on-line.

5.5   SEQUENTIAL ORGANIZATION

Sequential access

Sequential access to files, defined as retrieving each record in sequence, is applicable for large files which have high volatility and/or high activity in their processing runs. For example in a sequential file maintenance run, using magnetic tape as the medium, a new tape is written as a result of:

1.   Copying unchanged records from the old file to the new file
2.   Adding new records
3.   Deleting specified records

In this example, sequential access is used because it is the only practical means of access when magnetic tape is involved. Sections of the master file and the update file are shown in Figure 5–2. The master file contains details of inventory held, minimum balance, reorder quantity and unit cost. The transaction file consists of three possible transaction types:

1.   An addition to the master file
2.   A deletion from the master file
3.   A change to the master file

Old master file

Item number	Balance on hand	Minimum balance	Order quantity	Unit cost
1011	220	300	200	2.75
1326	1768	1500	750	13.52
2487	487	250	250	6.87
3215	52	25	50	3.28
3637	168	65	100	28.50

Transaction (update) file

Item number	Transaction code	Balance on hand	Minimum balance	Order quantity	Unit cost
1876	1	100	50	100	13.75
3215	2				
3637	3	188	65	100	28.50

Figure 5—2   Sequential file update example

To create the new master file, both the old file and the transaction file must be read, one record at a time, and the keys compared. The first update transaction (item number 1876) is greater than the first two item numbers on the old master file and consequently records 1011 and 1326 are copied without change on to the new file. On reading record 2487 from the old master, update record 1876 must be written on to the new file first and so on. Thus the process of key comparison creates the new file in correct sequence.

Direct access

Direct access is defined as the ability to go directly to any wanted record on the file without reading all previous records. Clearly, direct access is not always possible in a sequentially organized file. For example magnetic tape files do not allow for direct access, while magnetic disk-based files do. The most common method for achieving direct access to a sequentially organized file is by the use of indexes to indicate where a wanted record will be on that file. The concept of an index to a disk file is the same as that of an index to a library. When a book is required from a library, it is normal to look (say) in a title index to find the number allocated to that particular book, and from there its location on the library stack. In the same way, an indexed sequential file is one in which access to a sequential file which is stored on a direct access medium, is provided by the use of an index file. This index file contains the necessary information to indicate the location of the wanted record on the disk drive. This form of organization is called indexed sequential and it provides both direct and sequential access. It has proved to be one of the most popular forms of file organization in commercial data processing. Much commercial data has a natural sequence (such as product code sequence) and can be processed efficiently either by using that sequence (say in producing a stock report) or direct access (say in a low activity update run).

Figure 5–3 shows a simplified example of one method of arranging access to an indexed sequential file. In this case there is a two-level index, the highest level being the cylinder index and the other level the track index. The names refer to the concept of a cylinder and tracks on the disk surface. Data are organized as a sequential file on tracks 1, 2 and 3 of cylinder 7. The overflow area indicated on track 3 is space set aside for the addition of new records at a later time. To obtain direct access to record N, a search of the cylinder index shows that N is greater than F and less than S, and therefore the desired record is in cylinder 7. On searching the track index for cylinder 7 the key N is greater than K and less than P, and therefore the desired record is on track 2. A sequential search of track 2 then reveals the wanted record in the second block.

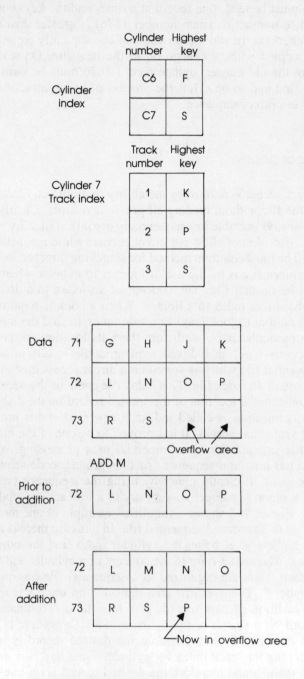

Figure 5—3   Indexed sequential example

When a new record is added to this index sequential file the physical sequence is maintained to the greatest extent possible. For example a record whose key is M will need to be added to track 2 but before this can be done space must be made as the track is currently fully occupied. This space is created by the software firstly moving the last record in the track (P) to the overflow area, then moving O to the last place and N to the second last place; thus allowing space for M to take its proper sequential slot. In this way the home track is still in sequence but some of the logical sequence for track 2 is now contained in the overflow area and, consequently, must be linked to track 2 in some way.

5.6 RANDOM ORGANIZATION

Unlike sequentially organized files, random files have no simple relationship between contiguous records. The method of organization is based on a computed relationship between the record key and the location at which the record is stored. thus, instead of using an index to indicate the record's address, a computation is made on the record's key to calculate an address. This process is illustrated in Figure 5–4.

One way of calculating an address is to divide the key by a prime number close to the number of blocks available on the disk for the file. The remainder of this division is the address. Thus if there are 200 blocks, divide the key by 199, and the remainder is used as the address of the record relative to the starting point of the file.

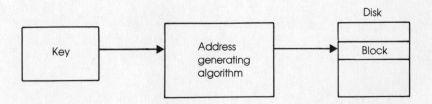

Figure 5—4   **Storing and retrieving records in a random file**

5.7 REVIEW QUESTIONS

1. (a) What is the difference between a master file and a transaction file?

   (b) Classify each of the following files (see Section 5.2):

   - File of customers
   - File of stock items
   - File of hours worked by employees during the week
   - File of hours worked by employees for year to date
   - File of items on back order
   - File of seat reservations by flight

2. What is the difference between sequential organization and random organization?

3. Explain the difference between file organization and file access.

4. (a) Define each of the following:

   - Volatility
   - Activity
   - Size

   (b) How do each of these affect the choice of file organization?

5. Discuss the effect of file storage media cost on file organization.

6. Outline the procedure used for direct access to an indexed sequential file.

7. Discuss the importance of files used in computer-based information systems.

8. In a computerized hardware store, what master files and transactions files would you expect?

9. Define the difference between files, records and data items.

6 Information Systems

6.1 INTRODUCTION

Information systems exist in a multitude of shapes and sizes with one common element: to satisfy the information needs of their users. For example a small motel operates a microcomputer to manage its room reservation and accounting system, while a railway authority uses a network of large computers to manage its fleet of rolling stock and freight business. It is the conjunction of user needs, available technology, and the skills of the analyst that determines the structure and shape of the system. A well-designed system exhibits the Bauhaus principle 'form. follows function'. In this chapter a number of categories of information systems are defined together with examples and, where appropriate, the main factors influencing choice of category.

Information systems can be categorized under each of four headings:

- The management level of the system users
- System response requirements
- System distribution
- System size

75

6.2 SYSTEM USERS

The principal system users are considered to be drawn from one of the three organization levels: operational, tactical and strategic. *Operational system users* are responsible for the day-to-day running of the organization and constitute the largest class of user. The bulk of information systems in most organizations are concerned with operational applications. Typical operational systems include:

- Order entry
- Inventory control
- Production scheduling and control
- Accounts receivable
- Accounts payable

As the operational applications handle the paper work for the day-to-day transactions of the organization, they tend to be characterized by:

- High volume of input and output
- Routine scheduled processing
- Updating and maintaining large files

Tactical system users generally are middle-management staff responsible for the medium-term planning and control of the organization. Typical tactical systems include:

- Marketing information
- Management information
- Production planning
- Budget control
- Vehicle routing
- Supply servicing and planning
- Project control

Tactical systems largely summarize the data maintained by the operational systems but may also have their own specifically designed input.

Strategic system users are senior management concerned with the overall performance and survival of their organizations in the medium to long term. For this purpose computer-derived summary reports, incorporating middle-management opinions, are perhaps the most common information system. Relatively few computer-based systems feed data directly to top management. However, systems which may be considered strategic include:

- Management information systems
- Decision support systems
- Strategic resource allocation systems
- Corporate models

Strategic-level managers influence lower level systems through the establishment of policies. For example customer credit limits in accounts receivable systems and inventory re-order points may be set by tactical-level managers on the basis of strategic-level considerations.

The triangle shown in Figure 6–1 is typically used to represent the relationships between the three levels of systems. The arrows between the levels represent the flow of data between the systems of each level. Most data flows up in increasingly summarized form. However, some data, as in the example given earlier, flows down.

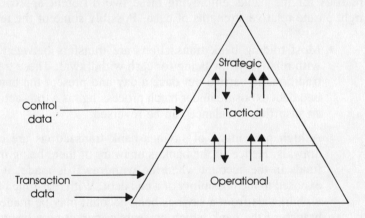

Figure 6–1   System and user levels in an organization

Typically, systems are not only located in one organizational level. These categories, like the other categories to be discussed in this chapter, represent 'pure' alternatives with an endless variety of systems possible between them. Thus, if a system is described as being at the operational level, this represents its primary focus and does not preclude tactical level functions.

6.3 SYSTEM RESPONSE

The response time of a system is the time delay between submitting or inputting a transaction for processing and completion of that processing being signalled by the production of output or another form of system response. Two aproaches significantly influence response time. The first, called *batch processing*, assembles transactions in a batch before carrying out processing. This results in a response time generally between several hours to several days, or even weeks. The second, called *real-time* or *transaction-oriented processing*, deals with each transaction as soon as it is submitted. Response times here are typically a few seconds. The following example is given to illustrate two commonly encountered batch and real-time systems.

Currently in Australia most trading banks operate batch processing of cheques and deposits to customer accounts with overnight running of the system to update customer balances. Thus the account balance is given for the close of trading on the previous day. Alternatively, most savings banks in Australia process customers' transactions using a real-time system to update the account balance both on the passbook and on the computer-based master file at the time of transaction. To conjecture reasons for the banks employing these two different approaches sheds light on the relative strengths of each. Possibly some of the reasons are:

- Most trading bank transactions are transfers between accounts, with relatively few being for cash withdrawal. These transfers are traditionally carried out once a day and present the bank with no bad debt exposure under batch processing, as a transaction against an insufficient balance can be reversed.

- A high percentage of savings bank transactions are cash withdrawals. Hence, if the bank is unaware of there being insufficient funds in the account when the withdrawal is made, it would be exposed to the possibility of a bad debt. With real-time processing, withdrawals from a savings bank account may be made from any branch of the bank which presents a marketing advantage.

- Presumably it was not thought important to allow trading bank customers to immediately withdraw funds from any branch of their bank.

If the input is submitted at a site remote from the computer, a real-time system requires the use of a data communications link to transmit the transaction data to the computer and to receive the return message. A batch processing system need not use data communications. The batch of transactions could be physically assembled and carried to the computer in either document or machine readable (e.g. magnetic disk or punch card) form. To continue our banking example, some trading banks magnetically encode cheques and deposits within the branch during the day and send these after work by courier to a regional processing centre.

Batch processing yields the following advantages over real-time processing:

- Better control of system integrity since control can be exercised at the transaction and batch level (see Chapter 7 for details)
- Easier system recovery in case of hardware or software failure. Because the batch system is seen in discrete program segments (e.g. the edit is completed before commencing the update) and at discrete times (with the output of each program being a file), re-runs due to a failure are comparatively easy to arrange
- Simple to synchronize transactions so that interference between them does not occur
- Cheaper to develop application software
- Cheaper to operate, since it can be run after normal hours

To summarize, the advantages of batch systems lie principally in their greater simplicity, their correspondingly lower development cost and their lower operating cost.

The advantages of real-time systems lie principally in their user-oriented features, namely:

- Immediate acknowledgement that the transaction has been successfully processed. For example a customer may telephone to change his address; a real-time system allows positive identification of his record and updating to be verified. A real-time reservation system is an example where acknowledgement is even more essential.
- Up-to-date status of important data. For example a real-time inventory system allows a salesman to check actual availability of stock, or a bank to establish the balance of an account.
- Immediate correction of edit and validation errors. Much time is wasted in batch systems locating transaction documents which have been listed on edit reports as failing a reasonableness test. If the error is signalled as the transaction is entered, it can generally be corrected immediately with far less trouble.

The term *on-line* refers to systems in which the input is physically connected by communication lines to the computer. An on-line system may be batch or real-time. A batch on-line system stores the transactions at the computer for later processing. However, the term on-line is also loosely used to designate real-time systems and it is wise when it is encountered to verify the intended meaning. For this reason it is probably as well to avoid its use.

6.4   SYSTEM DISTRIBUTION

In the early days of computing, around 10 to 20 years ago, computers tended to be geographically centralized in each computer-using organization. Perhaps the dominant factor in encouraging this practice was the economy of scale in computer manufacturing which enabled one large computer to be produced and sold much more cheaply than two smaller machines of equal total processing capacity. Due to a number of technical developments, this relationship no longer appears to hold, with the result that computer hardware cost is no longer the major driving force in deciding on the location and distribution of an organization's computers.

Today organizations still decide to centralize their computers though for reasons generally connected with data sharing and management control — both of which are easier to achieve with a centralized system. If data 'belongs' or is common to the organization as a whole rather than to a branch or division, transactions arising at any location are equally likely to require access to any of the stored data. Thus there is no logical way to divide up the database as it must be shared across the whole organization, and the obvious choice is to store it at one central site. An airline reservation system exhibits high data commonality and is run on a centralized system.

Management control is easier to achieve if all computers are physically located in the one area. Strict division of duties between programmers and operators, tight physical security, proper control of magnetic tape and disk libraries, auditing adherence to standards, control over system modifications and testing are all much simpler when there is only one computer facility. To visualize the difficulties of control with distributed computers, consider a branch operation of (say) 50 people with a small minicomputer located in the office area. Probably a single operator, doubling as a systems programmer and application programmer, would be employed to report to the accountant. Under these circumstances very few of the previously mentioned aspects of control could be easily achieved.

The alternative to a centralized system is a distributed computer system. A number of computers are located at sites physically remote from each other, generally with some form of data communication between them. This alternative moves the processing power to where the data is, and where the processing needs to be performed. Thus it finds application in organizations where data, and hence processing, can be easily segmented into a number of fairly independent clusters. Accounts receivable or inventory in a branch or division-oriented organization is an example. Distributed processing presents the following advantages over centralized processing:

- Reduced communication costs as data is processed where it arises
- Greater user identification with the system, leading to better and more creative use of the computer
- Greater user control over the system

The users are frequently the proponents of distributed computing in order to assert their control over processing. Electronic data processing (EDP) managers have generally opposed it on the grounds of reduced control of the system. Both users and EDP managers are correct in their assertions and centralization remains an issue for top management to resolve. In practice most larger organizations concurrently use a variety of centralized and distributed systems with the choice for each system made on the basis of data commonality, need for EDP control and need for user control.

6.5 SYSTEM SIZE

The geographical dispersion of computing power has been greatly assisted by the availability of computers in a continuous range of sizes, priced from many millions of dollars to a few hundred. Today an appropriate computer can be selected for almost any conceivable set of requirements. In this section the broad classes of computers are discussed, particularly as they influence the application systems designed for them.

Computers are broadly divided into three classes: mainframes, mini-computers and microcomputers. The edges of these classes are quite blurred with some machines equally termed a mainframe or a minicomputer. However the classification does serve to help differentiate the machines and their likely application.

A mainframe is the term used for a large machine generally costing in excess of around half a million dollars, and manufactured by the major computer corporations (e.g. IBM, Burroughs, Honeywell, ICL, Control Data, etc.). These computers perform the bulk of information handling for most large organizations. The name mainframe has confused historical origins and means nothing though it has stuck and serves to designate this type of machine. Mainframe computers are characterized by:

- Very fast processing speeds
- A long word size and large instruction set — so a lot of work is done in each machine cycle

- An extensive range of well-documented support software, e.g. file handling, database management software, communications network monitors, etc.

- Well-developed commercially oriented operating systems

Minicomputers typically cost from $20 000 to $500 000 and can perform most of (if not all) the same tasks as a mainframe, but at a slower rate. While widely used for commercial systems, they tend to suit smaller applications or applications that can easily be segmented into non-overlapping mutually independent systems and run on separate computers. For example, they are suited to a distributed computer configuration, where they have widespread use.

Microcomputers, the 'computer on a chip', cost from a few hundred dollars to $25 000 and are used as 'personal' computers for problem-solving in large organizations and for running commercial systems (e.g. debtors, creditors and inventory) in small businesses. Micros tend to be single user systems as opposed to mainframes and minis which service many users with the one computer.

Application software for microcomputers has to be cheaper than for large machines — users would not be happy to buy a computer for $10 000 and pay $100 000 for the application system. At first glance it may seem that the cost of analyzing, designing and implementing a system would be independent of the target computer. This is true in principle, but since a small computer can only run a small system, it is usual to find systems for small machines costing considerably less than systems for larger machines, even for the same general type of system.

One cannot compare (say) an inventory system developed in two man-months on a microcomputer with an inventory system developed in four man-years. Both systems may maintain stock records, but the larger system would no doubt service very many different kinds of products and cater for the often conflicting needs of a large user community. By contrast the small system would maintain records that perhaps were previously manually maintained by one man and produce reports to cater for the information needs of probably one or two users.

The cost of application systems for small computers is also reduced by use of standard packages where the development cost is amortized over a large number of sites. Packages have enjoyed much greater success in micro than in mainframe computers, possibly due to the economic necessity of re-using software and to the smaller organizational scope of the system — which makes changing the organization to fit the package an attractive alternative.

6.6   EXAMPLES OF INFORMATION SYSTEMS

Two systems are presented to help the reader understand the physical organization of a batch and a real-time system. In each case the system presented and the level of documentation has been greatly simplified to present the organization without the typical wealth of detail which may obscure the structure.

6.6.1   A Batch Payroll System

A payroll system at its simplest represents one of the easiest commercial systems to understand. However, practical payroll systems can be exceedingly complex due to the intricacy of wage and salary structures and the need to provide management reports for controlling personnel expenditure.

This case study presents a payroll system for a small computer consulting firm which is capable of supplying the following functions:

- Each employee is allocated to a department, has a job classification and a pay rate. All hours worked are paid at this constant rate per hour.

- The system is run each fortnight to produce:
    pay slips for employees (the cheques are written manually)
    a department expenditure report

- The system does not keep track of hours paid for holiday or sick time.

- Any extras paid are entered as a dollar figure on the input sheet.

There is one master file for the system, the employee master file. Its layout is given below.

Employee master file:
Employee number (key)
Name
Address
Job classification
Department number
Pay rate ($ per hour)

Current pay:
 week ending
 hours worked
 extras ($)
 gross pay
 tax deducted
 net pay
Year-to-date pay:
 week ending
 gross pay
 tax deducted
 net pay

Two inputs are required to run the system. The first, the *employee maintenance input*, informs the system of new hires, terminations, changes to existing employees and corrections to pay data. The latter data are needed in case of incorrect reporting of hours worked or pay rate, necessitating a manual change to the pay and then later correction of the history data held on the file. The format of this input form provides for any of these data to be included and requires the signature of the paymaster.

The second input form is the *time sheet* and contains:

Employee number
Hours worked
Week ending
Extras ($)

For control purposes a supervisor's signature is required on this form in addition to the employee's.

Figure 6–2 shows the system flowchart for the maintenance of the employee master file (e.g. new wage rates, new hires, terminations) in preparation for processing the time sheets. The employee master file is organized index sequentially. This permits the maintenance update to access the file directly, a logical choice given the low number of records accessed in this run.

Figure 6–3 shows the second run where the time sheets are keyed and edited before being sorted into employee number sequence. Transactions rejected by the edit program would usually be corrected and re-input before carrying out this sort. The update of the employee master file is carried out using sequential processing since all or almost all employee records will be accessed in this procedure. After updating the employee records, a pay slip is produced in duplicate. One copy goes to the employee with the manually written cheque, while the other copy is filed.

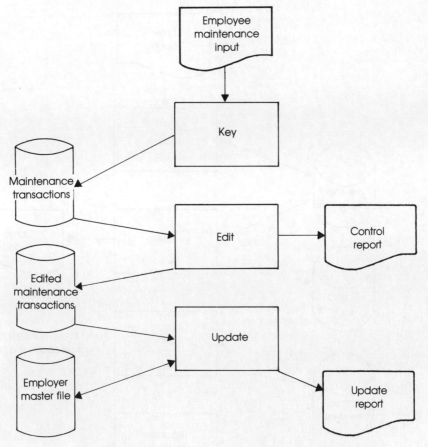

Figure 6—2   Maintain employee master file

Figure 6–4 shows the third run which produces the department cost report by first sorting an extract of the employee master file (only gross pay and hours are required) into department number sequence. The department cost report contains the following data:

> Department number
> Department name
> Period ending
> Current period:
>   hours
>   gross pay
> Year-to-date:
>   hours
>   gross pay

Total figures are given at the end of the report.

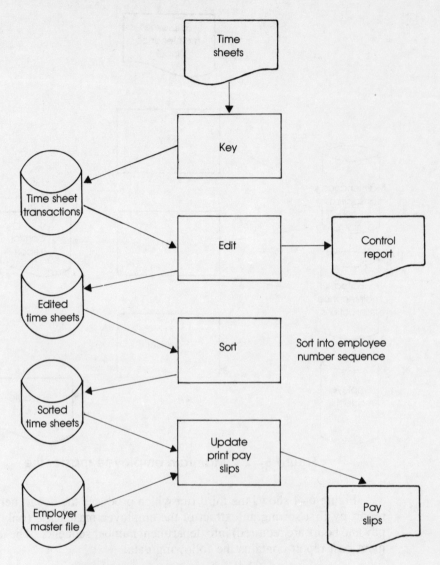

Figure 6—3 Enter and process time sheets

The department name file shown in Figure 6–4 contains:

Department number
Department name

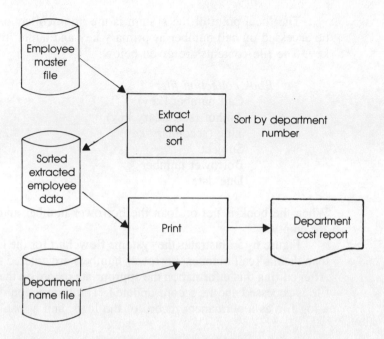

Figure 6—4   Produce department cost report

6.6.2   A Real-time Library Loan System

This case documents the system flowcharts and file description for a simple real-time library loan system. The system has the following functions:

- Loans are processed by recording (on the master file record of the book) the borrower number and the due date, and producing a loan slip with the book's call number and due date. This is placed inside the book to indicate it has been processed.

- Inquiry from a VDU enables the loan status of a book to be accessed. The book in question can be designated by either its call number or author and title.

- On demand a report can be produced of all overdue books.

The focal point of the system is the book collection file which can be accessed on call number as primary key and author/title as secondary key. The file contents are given below:

Book collection file:
Call number (key)
Author (secondary key)
Title (secondary key)
Subject
Borrower number
Due date

When the book is not on loan the borrower number and due date fields are empty.

Figure 6–5 illustrates the systems flowchart for the loan system. The loan details (call number, borrower number) are entered from a terminal. After editing this information the appropriate record on the book collection file is accessed and its record updated. The transaction data is written to a log file as a permanent record of the loan, and a loan slip produced.

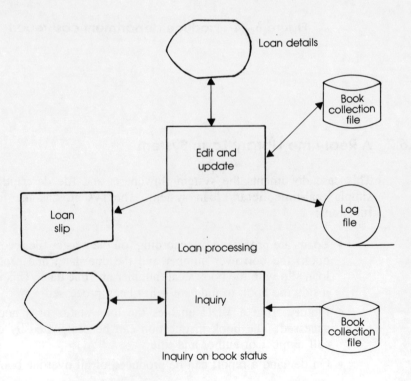

Figure 6—5 Loan processing and book inquiry

The on-demand production of the overdue book report is peformed by searching the book collection file for books with a due date earlier than the current date and printing these records. Figure 6–6 shows this flowchart.

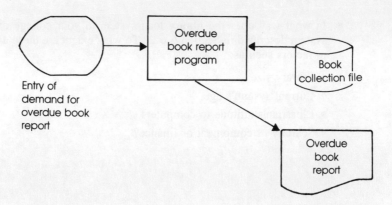

Entry of
demand for
overdue book
report

Overdue
book report
program

Book
collection file

Overdue
book
report

Figure 6–6   Overdue book reporting

6.7   REVIEW QUESTIONS

1. How would you expect systems used by strategic management to differ from those used by operational management?

2. Discuss the differences between batch processing and real-time processing.

3. In what way might a system combine batch and real-time processing? Give examples.

4. What advantages are offered by batch processing?

5. Two possible processing alternatives for a retailer could be described as:

   (a) Real-time, involving point-of-sale terminals to capture the sale of each item

   (b) Batch processing, in which sales information is derived from total cash and credit sales per department per day, with periodic stocktakes

   Compare the information available to management under these alternatives.

6. The Dean of the Business Faculty has just purchased a 'personal' computer. What applications would you recommend he run? In answering your question bear in mind the Dean's strategic management responsibility.

7. The batch payroll system in Section 6.6.1 has been simplified. In what ways would you expect this system to be expanded if it were to suit a practical purpose?

8. In what way does the library loan system in your college differ from the system described in Section 6.6.2? To what extent are these differences due to factors such as:

   • Library size
   • Current system's age
   • Librarian's attitude to computers
   • Scarcity of equipment or finance?

7 Controls

7.1 INTRODUCTION

Since any system, computer-based or manual, is subject to breakdown, error, fraud or sabotage, it is essential that a network of controls be established to try to prevent these problems occurring and, if they do occur, to rapidly detect and isolate them.

By far the most common form of errors are accidental omission or corruption of data. This includes such errors as mislaying data input documents, transposing digits when copying down a code, misreading badly formed characters and miskeying of data. Less common, but perhaps more worrying to management, are actions taken to manipulate the system to defraud the organization. Even after the data has been safely and correctly entered to the system it may become corrupted through machine malfunction. While this form of error is quite rare, it nevertheless needs to be guarded against.

Some controls are designed to prevent errors by ensuring an effective and professional approach is taken to the design, development and operation of a system. Other controls are focused on the prompt detection of errors by carrying out screening tests or duplicate operations.

Controls are usually classified into the five general categories listed, but a certain degree of overlap is present, making some forms of control difficult to classify.

91

Operational controls. These relate to the processing of data within the system and include:

- Input controls
- Processing controls
- File and database controls
- Output controls

Operational controls are developed in the course of designing the system and are therefore the responsibility of the systems analyst. This chapter will concentrate on principles associated with their design and cover briefly the other categories of controls.

External controls. These controls are exercised by groups external to the EDP function, such as the external audit firm or a staff control unit. Their purpose is to establish an independent opinion on the effectiveness and efficiency of the information processing activities of the organization. The controls are not reviewed in this text.

Administrative controls. These controls fall largely within the traditional management functions of the EDP department. They include such responsibilities as establishment of strategic plans, selection, hiring, training and allocation of staff, development and enforcement of standards.

Systems development controls. These controls relate to the standards for developing and documenting either a new system or maintenance to an existing system, and include the aspects of project planning and control and testing. For example the systems development lifecycle and top-down methodology presented in this text could form a control framework for the design activities.

Security controls. These controls include the physical measures used to control access to sensitive information and equipment, such as file libraries, computers and output. Physical control could include locked access, passwords and vetting of staff.

7.2   OPERATIONAL CONTROLS

Operational controls ensure the accuracy, completeness and authorization of all data processed, and consist of input, processing, file/database and output controls.

7.2.1   Input Controls

Control totals generally form the primary control mechanism for input transactions. In principle, a set of control totals should accompany the data all the way through the system. For a batch system the control is initially based on the batch. However, after editing, the batch identity may be lost through sorting together a number of batches. Hence the control would at this stage of processing be based on the group of batches. For a transaction processing real-time system, since each transaction is processed as it is input, greater care must be exercised in designing other forms of control for the individual transaction; these are discussed later under programmed controls. Additionally the processed transactions can be batched and a batch control total checked, *ex post,* with the processed data.

The total used for operational control frequently takes the form of a hash total. This is the total of a sequence of numbers regardless of the units. For example, the transaction values given below have a hash total of 62.92:

> 15 gallons
> 32.67 millilitres
> 15.25 kilograms

As batch hash totals must be manually prepared, the designer should resist the temptation to ask for hash totals to control every field of input — a hash total is to be used to control important fields only, particularly those where no subsequent checking can be done. For example, a customer code on a cash received advice would not normally be controlled with a batch hash total. If the code is miskeyed it most likely will be discovered through a mismatch with the customer master file in subsequent processing.

The precise design of input controls obviously depends on the nature of the system, the degree of protection required, the system's external environment and the nature of the organization. The following points are generally desirable:

- Each input batch should normally be less than about 50 documents. This is dependent, of course, on the nature of the input, with large and complex documents needing a smaller batch size. The objective is to facilitate the tracing of input errors; large batch sizes make this task more difficult. After the edit stage the input batches in a computer run are, for control purposes, merged into one batch.
- All documents sent for keying need to be accompanied by a batch header slip which contains (a) batch sequence number, date and type; (b) a count of all documents in the batch; and (c) hash totals

of quantities and values of key fields which should be carefully controlled. (Examples are the quantity field of an order and the cash received field for an accounts receivable system.)

- With advanced data entry equipment the control totals can be checked after the keying of each batch. But more generally they must wait until being input to the edit program to be checked.

- At the end of each batch in the computer input run, the counts and totals are printed; alternatively they may be stored and printed as a table after reading all the batches. As an additional security measure the control counts and totals themselves may also be input and checked automatically against the figures accumulated by the computer. The details of any discrepancy are then not only printed but, if desired, the computer is programmed to require special operator action to be carried out before the run can proceed. There is then less chance of ignoring a discrepancy — deliberately or unintentionally.

- After each program run the document counts and hash totals are again printed including any new calculated figures created as a result of the processing, and for which control is required. The latter are especially important if they relate to data that are used in the subsequent runs.

- Amendments to or updating of a master file should be accompanied by a simple analysis (control account) of the master file before and after the run. This includes the numbers of records and any suitable data field totals. The analyses are dated and retained on hard copy as a visible history of the file, and they are also held on the file itself, generally in a special record at the beginning or end of the file.

- A record of the computer runs and tape and disk file identification should be printed on the computer operator's terminal (normally called the console). This record, termed the console log, needs to be preserved so that the precise sequence and nature of the computer jobs run can be checked. This can be used to identify if errors occurred in running the job (e.g. wrong job sequence, wrong files mounted) or unauthorized computing was carried out (e.g. a disgruntled programmer running the payroll master file to increase his rate of pay).

The EDP department and the user department which submits the data need to share the responsibility for data controls, but ultimately control responsibility must rest with the user department. Sufficient information must be manually maintained by the user department to satisfy it that all data submitted for processing has indeed been processed. The biggest source of weakness in many installations is the handling of reject documents from the edit program. When only a few transactions are rejected from a batch by the edit program the normal procedure is for the batch total to be adjusted for the rejected transactions and the processing run continued. (If too many transactions are rejected the whole batch is rejected and returned.) The rejected transactions are married to the source documents and returned to the originating department to be corrected. The potential source of weakness here is that as far as the EDP department is concerned their data controls indicate balance, and the returned documents are no longer their responsibility. For the originating department there is a temptation not to effectively control these rejections as they are already included in a previous day's totals. Cases have been known where a search through the desk of a clerk following his quitting have unearthed many rejected unprocessed documents lying at the back of the bottom drawer and dated months previously. One of the most compelling reasons for organizations to go on-line for data entry and editing is to avoid the costly delays and problems that occur in correcting and controlling rejected transactions.

Other forms of input controls are listed below with typical examples of their use:

- Sequence checks: if batches are numbered sequentially a sequence check enables a missing batch to be identified.
- Input authorization: certain classes of transactions such as credit rates or large purchases may require special authorization by a supervisor from a separate terminal.
- Encoding verification: for a real-time system the correct key entry of a code (say a product or customer code) can be visually verified by the operator checking that the description or name displayed on the screen agrees with the code entered. The response displayed on the screen is obtained by using the entered code to access the relevant record on the master file.

7.2.2 Processing Controls

Processing controls consist of a variety of procedures incorporated into the application program code that help ensure only valid data is processed. Processing controls are generally concerned with screening input and are closely allied to the input controls. Indeed many input controls (e.g. batch total controls) rely on procedures in the application program to function. However, processing controls refer to those procedures which function independently of externally provided control input, and consist of programs designed to edit and validate input.

All data first entering a computer system need to be edited. The objective is to detect by means of logic checks every conceivable form of error. Hence every check that can be applied to the data should be. A rich source of ideas on editing data can frequently be gained from the clerks and supervisors responsible for manually creating and processing (if the previous system was manual) the data, i.e. the checks they applied to determine if the data looked acceptable. Some of the typical edit checks applied to data include:

- Limit checks: these are predetermined limits that all values must fall between. For example quantity ordered may need to be in the range 1 to 300. In a real-time system the limit checks may be set more tightly with a supervisor or operator over-ride.
- Combination check: a test applied to a number of fields simultaneously. For instance a transaction may have a price and a quantity field; a combination check may be set up on the product of price and quantity.
- Restricted value check: a test applied to a field which may be known to take on only a certain number of values. For example a field indicating sex may be only M or F, or 1 or 2.
- Format check: all input documents have a defined format which needs to be checked. For example numerics in numeric fields, alphas in alpha fields, and field sizes correct and in the right location.
- Relationship checks: frequently a certain code in one field may require a certain range of codes in another field. For example the code F under 'sex' will limit the range of codes under 'title' to exclude the titles MR and MASTER.
- Check digit: this is the name given to a digit in a code which is mathematically derivable from the other digits by a formula. Application of the check digit formula and comparison with the check digit can tell us whether the code is legitimate. This can prevent the further processing of a transaction where the code is meaningless due to either miscoding, transposition of two digits, misreading the input document, etc. A number of formulae are used for calculating a check digit. One of the most common schemes is called 'the modules 11 check digit'. The method of computing it is:

  1. Multiply each digit in the code number by its weight. The weight for the least significant digit is 2, the next least significant digit is 3, and so on.
  2. Add together the above products.

3. Divide this sum by 11.
4. If the remainder is 0, the check digit is also 0. If the remainder is not 0, subtract it from 11 to give the check digit. A check digit of 10 is usually written as 'x'.

The reader should verify that the check digit in the code 389 is 1.

These edit checks refer to internal checks of the data by recourse to pre-defined logic rules. Validation checks, on the other hand, refer to checks based on external references to file information. For example a transaction to delete a customer from the master file may be validated (before attempting to process the transaction) by checking to make sure that the customer is actually on the file. Information that may be validated includes codes, names (in a real-time system the name displayed on the VDU allows manual validation in response to a code being input) and status condition.

Whereas the edit checks are compulsory on all systems, the validation check needs to be decided on the basis of judgement in each individual situation. If no validation program or module is included the error will be come apparent at update time. However, dealing with the problem then is a little more difficult than clearing all data prior to update. In this case, the cost of both the development and the operation of the validation process have to be weighed against the problems caused by occasional errors being detected and rejected at update time.

7.2.3 File and Database Controls

File and database controls are concerned with safeguarding the stored computer-based data of the organization. Since these data represent the foundation of the organization's information system, they must be safe-guarded with great care.

File and database controls rely on the computer through the use of either standard utilities (such as a file copy) or specially designed code in the application system. The principal types of procedural controls are:

- Establishment of back-up copies of files
- Integrity controls

Establishing back-up copies of files may be simply done in sequential tape file processing using the grandfather, father and son procedure. In this procedure three versions of the file are kept at any one time. File 1 (father) in update cycle 1 produces file 2 (son). In update cycle 2, file 2 (now a father) produces file 3 (now the son). Update cycle 2 converts file 1, previously the father, to a grandfather. See Figure 7–1 for the flow-chart of the procedure.

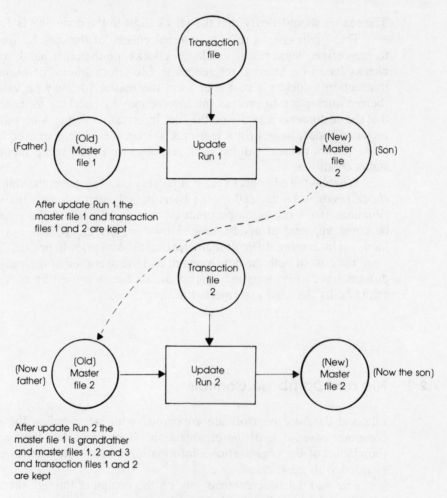

After update Run 1 the master file 1 and transaction files 1 and 2 are kept

After update Run 2 the master file 1 is grandfather and master files 1, 2 and 3 and transaction files 1 and 2 are kept

Figure 7 —1   Grandfather, father and son procedure

Since the old transaction files are also kept, this approach enables recovery from the loss of both the present master file and the previous version.

With direct access storage devices (DASD), since updating takes place *in situ*, file back-up is accomplished by periodically copying the whole file or whole disk (or both) and storing the back-up copy. Naturally the transaction file contents need to be also stored to enable an old master file back-up copy to be rolled forward by reprocessing the transactions.

File integrity controls, as the name suggests, are concerned with safeguarding the accuracy and integrity of the stored data and permitting reconciliation with the input data controls. Stored data integrity can be jeopardized either by human or computer hardware error or by fraud.

As the whole file is passed in each processing run in sequential file processing, it is comparatively simple to compute file controls by counting the number of records on the file and summing key fields. These totals can then be compared with calculated totals prepared on the basis of total transactions input plus total on the file at the end of the previous processing run. These 'run to run' controls afford a very high degree of file integrity, particularly when used in conjunction with a grandfather, father and son method of file version retention.

As an example consider a simple accounts receivable file which has information on invoices, cash received and balance due. After each update run the data could be printed as in Figure 7–2. The data marked with an asterisk would normally be held on the file, and the control reports checked to ensure the carried-forward (c/f) figure of one run equalled the brought-forward (b/f) figure of the next run.

CONTROL REPORT XYZ 99/99/99   Page 1
                   CALCULATED TOTALS
b/f Balance due*   21 000.00
     add invoices   8 000.00
     deduct cash received   10 000.00
c/f Balance due*   19 000.00
                   SUMMED TOTALS
b/f Balance due   21 000.00
c/f Balance due   19 000.00

Figure 7–2   File control report

With direct access files it is uneconomical to read through the entire file after each updating run to sum the key fields. So only the records updated are included in the addition and the assumption is made that the other records have not changed. Every so often the file must be checked to ensure the actual sum of the fields equals the calculated sum. This is normally done each time the file is backed-up when it must be read in its entirety in order to copy it.

7.2.4   Output Controls

The major concerns are dissemination of output and the verification of the necessary relationships between input and output. Systems should exist, therefore, to ensure that output is distributed to authorized persons only, and that the data contained in that output has been reconciled with the various controls established over input and files. This is normally done by a person reviewing the control totals and establishing their validity.

All computer-printed output must be clearly identifiable in terms of its exact particularity. This entails the precise labelling of every sheet of print with appropriate information such as captions, date and page number.

7.3   ADMINISTRATIVE CONTROLS

Administrative controls encompass the normal administrative and management functions associated with the running of the computer environment. They include:

- Establishment of plans
- Control over staff
- Division of responsibility
- Stand-by facilities

7.3.1   Planning

With long lead times in equipment acquisition and software development, and scarce human resources, the effective functioning of an EDP department requires a thorough on-going commitment to planning at the strategic and tactical levels (project planning is included in system development controls). The strategic plan gives a broad outline of the objectives and direction of the EDP department over the long term (e.g. five years). As this department operates as a service resource for the organization, its plans must necessarily be very closely dovetailed to the information needs of the organization as revealed in the general strategic plan.

Some of the issues that may be addressed in the strategic plan include:

- Future development of the data communications network
- Broad areas of new project work
- Distribution of systems and databases

- Database development and language
- Systems and language standards
- Computer upgrade path
- Personnel growth

Tactical planning involves the consideration of the medium term (generally one year) and is concerned with such questions as:

- Project plans
- New hardware implementation
- Personnel training plans

7.3.2 Control Over Staff

Aside from the question of division of responsibility, control over personnel is achieved through activities such as:

- Recording of all transactions with time stamp, terminal number and employee number
- Establishing a schedule of operations so that any unusual runs are noticeable
- Authorization, where possible, of sensitive transactions
- Controlling output distribution
- Defining personnel duties, responsibilities and access restrictions
- Enforcing vacations and, where possible, rotating duties
- Setting standard procedures for activities

7.3.3 Division of Responsibility

A normal control requirement to provide a cross-check on the accuracy and propriety of systems is the separation of the planning, design and operations activities of the computer department. Given the use of in-house computers by small organizations, this is not always possible. In this case the organization runs the risk, for example, of unauthorized changes being made to software, or of modifications being made while systems are running. Ideally, EDP management, the user, operations, system design and programming, the librarian and the software assurance functions should all be separated.

For example in a payroll system the following people could play a role with the numbers indicating the time sequence:

1. *Supervisors* sign time sheets
2. *Paymaster* calculate time sheet batch totals
3. *Data input clerk* enters time sheet batch in batch book
4. *Computer operators* schedule keying of data and running of system
5. *Output dispersal clerk* reconciles input controls to output controls on pay sheets and enters dispersal in batch book
6. *Computer file clerk* checks file controls reconcile with input controls
7. *Paymaster* checks paysheet controls with time sheet controls

7.3.4   Stand-by Facilities

These controls are closely related to back-up, but cover hardware. Consideration should be given to the need for, and the availability of, an alternative facility which could carry out critical processing for a period of time, in the event of an extended breakdown of any part of the computer machinery. This is normally achieved by establishing a reciprocal relationship with another organization using the same computer configuration.

7.4   SYSTEM DEVELOPMENT CONTROLS

The environment in which systems are developed will have a considerable impact on the quality of the systems produced, and consequently adequate control over this development process is of concern to all organizational participants. The areas of concern include:

- Documentation
- Project management
- Testing

7.4.1   Documentation

Controls in this area aim to provide a secure record of the system design, programming and maintenance activities, as well as user procedures and recovery procedures for each system. This involves setting adequate standards for documentation, and ensuring that access to that documentation is controlled. One of the major problems faced is maintaining the accuracy of the documentation as systems are modified over the years.

7.4.2 Project Management

The project management environment does much to determine the quality of systems and the efficiency of their development. Sometimes the audit objectives embrace efficiency concerns and thus the auditor is concerned that adequate feasibility studies are carried out, and that user participation is sufficient to ensure that developments are in line with user needs. Further mechanisms should exist to allow resource (e.g. analysts, programmers and computer time) budgeting, allocation and review.

7.4.3 Testing

The procedures for testing new and modified systems should aim to minimize the likelihood of system faults in the operational environment. This means that standards for testing procedures should be established, with the aim of the tests being to uncover any improper functioning in the system.

Testing of computer software may be carried out in a number of ways. One commonly-used aproach is to start testing at the module level and progress to the program level, the sub-system level and finally the system level. This testing approach is called bottom-up and is described below.

Module testing. A module is a contiguous segment of code that (perhaps with a test harness to pass and receive values) can be tested independently. Module testing may be omitted where the programs are relatively simple and comprise small non-complex modules.

Program testing. When a program has been coded it must be tested to verify it performs as specified. This testing and the attendant correction of errors usually takes longer than the design and coding of the program.

Sub-system testing. The system can be thought of as a number of sub-systems and each of these can be independently tested.

System testing. When finally all the pieces are ready, the whole system is tested — initially with test data and then with live data.

7.5 SECURITY CONTROLS

Physical asset security is concerned not only with the system of protection and insurance for the hardware, but with the software which makes up the computing capability. Thus consideration should be given to the security of and access to application programs, data files, system software (including utilities) and hardware. Some examples of the nature of these controls are:

- Authorization for access to application programs and files
- Access restrictions on the computer room
- Control over general system software usage commensurate with its capability, through misuse, to cause loss or corruption of data
- Fire detection facilities

The purposes of physical asset controls are to ensure that:

- The files and programs are secure from unauthorized access and alteration when on-line
- Utilities are used only for legitimate purposes
- The going concern assumption is not endangered by destruction or modification of either hardware or software

In any configuration involving terminals, procedures are needed to ensure that access to the system is gained for legitimate purposes only. Passwords or some equivalent should be used to restrict access to the system or parts of it, and the terminal itself may be restricted to certain functions or data needed by its users. Further control is provided by automatic logging of all terminal usage. Possible attempts to use the system illegally can be highlighted by a review of the log together with an exception report on invalid use (such as failed access attempts). At the data file and data field level, security access facilities in the file management system provide yet another layer of control.

7.6 REVIEW QUESTIONS

1. Name the various control categories and briefly distinguish between the objectives of each.

2. Refer to the batch payroll system in Section 6.6.1.

(a) Determine the edit rules that are desirable for the edit program of Figure 6–3.

(b) Design the control report produced by the edit program in Figure 6–3.

(c) What file control and output control data should be produced after the completion of the update and print pay slips run (Figure 6–3)?

(d) What backup procedures would you recommend for this system?

3. What administrative and security controls are particularly important in a payroll application?

4. The form shown below is used to transfer stock between different cost centres within the Kenso Manufacturing Company. A number of items could be transferred using the one form. The organization operates a batch inventory control system, processing stock transfers daily.

Design a batch control header slip for this application.

KENSO MANUFACTURING COMPANY

No....................

Date...............................

RECEIVING COST CENTRE NO.......................................

SUPPLYING COST CENTRE NO.......................................

Product code	Description	Quantity	Unit of measure
	HASH TOTAL		

Issued by	Goods received by

5. Discuss the controls you think are necessary for the edit and update process (Figure 6–5) for the library loan system described in Section 6.6.2.

6. Calculate the modules 11 check digit for the number 138 570.

7. List the possible edit checks and illustrate their use through suitable examples drawn from business systems.

8. Why is it not normally considered necessary to have a control total on a key field such as stock number?

9. Explain the grandfather, father and son technique for updating a file.

10. What controls would you recommend in a stock control system to ensure that the same transactions were not entered twice?

11. Investigate and describe the file controls used in a data processing centre known to you.

8 Systems Analysis

8.1 INTRODUCTION

Systems analysis is the study of a system's problem including the identification and analysis of various alternative solutions. In system design the particular alternative solution that has been decided on is considered in detail and an appropriate detailed design produced ready for programming. This distinction between analysis and design is not rigid but is suggestive of the primary focus of these two phases of systems development activity. Aspects of analysis and design work are performed throughout both phases. Thus the analysis phase involves a component of design in the creation of alternative solutions while the design phase involves analysis, for example, in detailing output specifications. Although this chapter specifically examines systems analysis, it initially presents an overview of both analysis and design.

The detailed analysis and design of an application system involves the use of a range of skills including interviewing, fact finding and negotiating political compromise, and the use of tools such as those covered in Chapters 2, 3 and 4 for the collection, documentation and analysis of a bewildering array of data. Finally, the task requires the preparation and documentation of a number of alternative designs. The system development process is not a straightforward progression through a number of defined stages, but is iterative in nature as the activities and work carried out in earlier stages are modified to reflect one's improved understanding of areas that were not sufficiently explored earlier.

8.1.1 A Model of Systems Analysis and Design

In undertaking a system development project the analyst needs to be armed with a good methodology that will lead him through the application of the relevant tools and enable control of the project to be maintained. It is now generally believed desirable for the methodology to provide a structured top-down design approach. By structured is meant an orderly, systematic, step-by-step approach which if followed will yield a solution to the problem.

A top-down design approach first identifies the major functions to be accomplished, then proceeds to identify the next level sub-functions required to implement the major functions and so on. This leads to a tree structure of functions with each function defined more explicitly in lower levels of the tree.

Perhaps the top-down approach may be seen more clearly by considering its mirror image, a bottom-up approach. Here, the lowest-level functions (perhaps the tough knotty little algorithms) are first defined and then the upper-level integrating functions are defined.

A top-down methodology enables the designer never to lose sight of the major functions of the system and ensures the lower level sub-functions can be integrated into the higher level units. With a bottom-up design, as integration is considered late in the design, the lower level modules designed first may not fit together. A cardinal tenet of top-down design is to preserve flexibility by only making those decisions which are necessary for the immediate job at hand.

The top-down design methodology is demonstrated using a block hierarchy diagram to illustrate the functions required in a system to look after club membership records. The overall function 'maintain and process club membership records' as the objective of the system occupies the top box as shown in Figure 8–1. The functions required to support this objective can be listed as follows:

- Add new member to file
- Delete existing members
- Modify records of existing members
- Bill and collect membership dues
- Produce membership lists or mailing stickers in alphabetical or postcode sequence
- Calculate membership statistics

These functions are shown on the second level of Figure 8–1 with the three functions add, delete and modify member records grouped together under the one heading — maintain records.

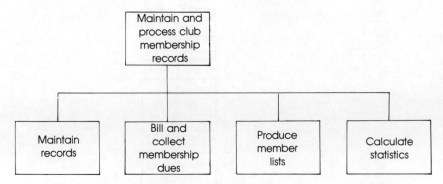

Figure 8—1   Top-down analysis of a club membership system

The top-down analysis now proceeds to identify and describe the functions required for each of these second level functions. For example the function 'bill and collect membership dues' may be seen to consist of:

- Preparing bills
- Updating members' accounts for money due
- Updating members' accounts when money received
- Determining accounts not paid and sending overdue notices.

In this way the top-down design proceeds progressively from the general to the detailed.

In the particular methodology presented in this and the following chapters, a block hierarchy diagram is not drawn but the essence of the top-down approach is preserved. Look for it as you read on.

Figure 8–2 illustrates the major activities in the structured top-down methodology for systems analysis and design which is presented in this and the two following chapters. It can be seen that there are five major steps in this model:

- The study of the current physical system
- The study of the current logical system
- The identification and study of design alternatives
- The design of the new logical system
- The design of the new physical system

The content of these steps, as included in the Figure 8–7 framework, will be spelled out in greater detail later but may be briefly described as follows.

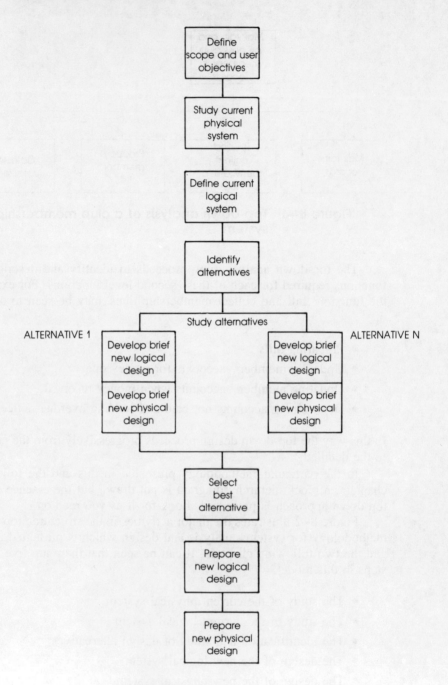

Figure 8—2   Steps in the analysis and design of a new system

Study current physical system. The starting point of analysis is the current physical system. The word physical denotes that all relevant details of the actual system and its environment are captured and studied. This includes, for example, the kind of physical data filing and retrieval systems used and the attitudes and perhaps cognitive styles of the user. The purpose of studying the current physical system is both to discover the present system's strengths and weaknesses and to develop an understanding of the system requirements.

Study current logical system. At the physical systems level we are interested in the full system — what is being done and how the processing is accomplished. The current logical system is the 'what' of the current system. It distills the mass of detail gathered in studying the physical system to obtain the underlying logic of what is happening. At this basic level there is frequently little difference between a manual and a computer-based system. Thus the product of the logical systems study activity is a specification for the procedures that need to be carried out to fulfil the same business objectives as the present system. As well as forming a foundation for the systems design, this stage enables the current logic to be analyzed for weaknesses.

Identification and study of design alternatives. A number of system alternatives are identified which meet the user's objectives and require-ments (e.g. to correct the inadequacies of the current system and add new features). The physical system implications of each alternative are spelled out in sufficient detail to enable technical and economic feasibility to be established and a decision made on the preferred alternative. This estab-lishes the target physical environment including computers, input/output media characteristics, and communications and database facilities.

New logical system design. Here the user's objectives for changes to the system functions are incorporated into the design. The changes may range from additional reported information to whole new operational pro-cedures that were previously uneconomical (given the then current systems technology).

New physical system design. The new logical design communicates what functions are to be carried out; the physical system design shows how they will be implemented. Thus the physical design translates and packages the logical design for the chosen technology: computer proces-sor, database or file management system, communications network access method, local terminal intelligence, peripheral storage capacity, etc.

Some of the top-down characteristics of this design model are:

- The top level functions (the definitions of the system, its scope and objectives) are settled first

- The logical functions of the current system which will form the skeleton for the new system are defined before considering new features
- Physical implementation decisions are delayed until the logical specification of the system is complete

8.1.2 Tasks in the Analysis and Design of an Information System

Initiating the system study

Ideally, systems analysis is preceded by a request incorporated in terms of reference for the study, laying out specifically what the objectives of the systems analysis exercise are, the scope of the study, and management's perceptions of the problems present and an expression of the benefits that they hope to realize. In some cases, systems analysis is preceded by a less formal document, or maybe only a verbal communication in which management expresses the feeling that a certain area of the business may lend itself to study because they suspect better technology or better systems design may lead to some cost savings or other desired benefits. Indeed, apart from projects involving minor modifications to an existing system, it is rare for a user to be very definite about what needs to be done, or even what the scope of the systems study should be. Whatever initiates the systems analysis study, it is essential that the management in the user area are in support of the work and that they believe there is a possibility of improving on the present system.

Performing the analysis

Following the receipt of the terms of reference, the current system is studied at the physical and then at the logical level. This thorough study of the current system and its environment permits the analyst to work effectively with the user in defining the user's objectives and drawing where appropriate from the organization's strategic plan in identifying alternative solutions. This is the creative heart of the analysis exercise and can only be well executed if both user and analyst have sufficient knowledge of each other's area and mutual trust to be able to communicate freely and work as a team. All too often distrust, excessive pride, secrecy and a lack of understanding of the need for team work prevent this user/

analyst cooperation ever emerging. Some techniques available to asist in building this level of mutual cooperation are dealt with in Section 8.2.

Recommending a solution — the feasibility study

In the feasibility study segment each of the proposed alternative solutions are examined by first developing a brief new logical design and then a brief new physical design. This would omit much of the detail and be principally directed at establishing the size, type and cost of the required hardware resources including files storage, computer processors, communications and input/output devices. These steps are required to examine the economic and technical feasibility of the proposal. When all alternatives have been studied, the selection can take place. The cost/benefit data will form a significant contribution to the decision, but a number of subjective areas will also be very important — highlighting the extent to which the user needs to contribute to this stage.

Preparing the design

Following the selection of the most attractive alternative solution the analyst sets to work on the system's design. Many facets of the system will need to be defined precisely before the detailed design specification can be developed. Chapter 9 considers this design stage in detail.

8.2 FORMING A CREATIVE TEAM

A number of studies have been carried out to identify the most important success factors of a computer-based information system. Some of these studies have examined failed systems to determine what went wrong (e.g. Lucas 1975b) and others have sought the opinions of computer professionals and managers (e.g. Carter et al. 1975). A common conclusion is that the identification of the information needs of management and the involvement of the user in the project are two of the most important factors associated with successful information systems development. These two factors are in reality mutually inclusive and dependent — effective user involvement ensures good requirements definition while without user involvement it is almost impossible to define requirements correctly. Thus, setting up the team should, desirably, be the first step undertaken in a project.

Both the user and the systems analyst need to prepare themselves before they are able to work together effectively. The user needs to become familiar with computer systems and the systems analyst needs to gain knowledge of the user's job. If the user is not sufficiently prepared for his role, he will not be able to approach creatively the task of defining requirements and identifying possible alternative solutions. For example an inventory clerk whose only experience is working with a manual system which updates inventory cards once a week may feel very bold in defining a requirement for an update every second day. A person's expectations and requirements are greatly moulded by their understanding of what is feasibly available from an economic and technical standpoint. Another example to illustrate an error on the side of excessive expectations may be a science-fiction enthusiast inventory clerk who lists a requirement for a system that has two-way man-computer voice communication capability for customer ordering and for inventory interrogation and reporting. But in this case, at least, the user demonstrates a motivation to think creatively about his job and the role that computer technology can play in assisting a better job to be done. This is the user frame of mind necessary for successful requirements definition. The term 'unfreezing the user' has been used to denote this task of preparing the user for requirements definition by exposing him to a number of exciting alternative ways in which computer systems can assist him.

One way that unfreezing may be accomplished is for the user to visit organizations with alternative systems implementations, or view technology in an area where it is relatively advanced. For example viewing an advanced colour graphics-based computer design system may stimulate managers to think of alternative ways of receiving information other than through printed reports. Each project will need its own carefully thought out approach to unfreezing the user. Obviously some users are well aware of current technology and alternative systems' approaches and no preliminary work needs to be done. However, in the case where the user is firmly entrenched in the current manual system the price often paid for not 'unfreezing' the user is a computer system that is a copy of the manual system and later on a stream of enhancement requests as the user is dissatisfied with the low level of the system, possibly as a consequence of seeing other, better systems.

Systems analysts have typically shied away from exposing users to advanced systems, possibly fearing loss of control of the management of users' expectations and requirements. The systems analyst needs to recognize the legitimate and necessary role of the user in defining requirements and the proper role of the systems analyst as catalyst, organizer of information, project secretary and technical advisor.

The systems analyst needs to learn about the user's job environment, the current system and its business impact before being able to creatively interact with the user. This is accomplished during the steps in Figure 8–2 called 'Study current physical system' and 'Define current logical system'.

8.3 DEFINING THE SYSTEM'S SCOPE AND OBJECTIVES

The first step of the system team is to establish and obtain agreement on the scope and objectives of the systems study. This is an important exercise and embraces both political and technical issues. If the problem area is wholly under the control of one user, the political aspect is likely to be minor, but if the area overlaps a number of responsibility areas or is to be a common system for a number of users, the political aspect is likely to be quite dominant. As computer systems always bring change, there are usually a number of winners and losers in the power stakes. The losers may be, for example, a department manager who, following the implementation of a system, will have his staff level reduced, or an accountant who previously had the responsibility of preparing the management information reports which will now be prepared on the computer. The winner is often seen as being the EDP department — one reason for the suspicion with which this department is viewed by users. The failure on the part of the systems analyst to recognize the political nature of the definition of the system's scope may lead to either significant problems in obtaining agreement on the system's scope or, worse still, significant problems in securing cooperation with the implementation of the completed system.

The system's scope needs to be broad initially and narrowed down as successive areas are omitted from the study. This will ensure that the scope eventually selected exists within a well-understood environment. Some points that should be considered in arriving at the proper scope for study are:

1. Resist the temptation to excessively widen the system boundary. Systems analysts are trained to have a systems or holistic view, which may lead them to err on the side of defining too large a scope for the system. An example of this may be an organization which has requested a new inventory system. The analyst defining the scope of the system may, on examining the order entry and invoicing, feel this application also needs to be included as it involves the entry of data to the inventory system. After checking on the receipt of purchased goods, the accounts payable system may also be included in the scope. Since many small problems are much easier to solve than one large one, it is simpler to carry

out a number of systems analysis studies integrating each one with the work done previously, than to work on one large project encompassing them all at one time. Wherever possible, it should be the aim of the systems analyst to de-couple systems so that they can operate, as far as possible, independent of other systems. In this way, errors and omissions can be localized in their impact and subsequent maintenance work is easier. This does not mean that systems should not be integrated, but rather that the systems interfaces be tightly managed to ensure they are simple and pass on the minimum necesary data.

2. The system's scope and objectives should be sized so the system is able to be implemented before the user's enthusiasm declines to such a point that success is jeopardized. For some organizations this may be around 6 months.

3. Hooks need to be provided in the system for linkage to future systems so that a narrow scope does not lead to future problems with system integration. This means that the system's interface requirements need to be included in the project scope.

4. The scope should be chosen so that the potential benefits are maximized.

The user's objectives for the systems analysis study may be initially quite vague (e.g. investigate our inventory system and see what the problem is — we keep running out of stock of some items and being over-stocked in others). Alternatively, more experienced users may be able to give quite specific objectives to the system's study team. Specific objectives enable the team to be more directed both as regards defining the system's scope and in planning the analysis. However, the right balance needs to be struck — if the objectives are overly detailed it may unne-cessarily restrict the freedom of the team to explore creative solutions.

Concurrently with defining the system's scope and user-oriented objectives, the technical objectives need to be decided on. These objectives are additional to the user's performance and functional objectives. They need to be explicit as they influence design decisions and cost/benefit tradeoff decisions that are constantly considered by the members of the project team during the design stage of the project. The objectives can be grouped under the following eight headings:

- Flexibility and maintainability
- Schedule and cost
- Efficiency
- Integration

- Security
- Reliability
- Portability
- Simplicity

However, these objectives cannot be considered singly because many overlap and can be in conflict with one another. Consequently, a balance needs to be struck depending on their relative importance in a given situation. An obvious example would be a conflict between the goal of providing portability of the design to different system software and hardware environments and the goal of efficiency when running in a given environment.

8.3.1 Flexibility and Maintainability

Flexibility and maintainability have been grouped together as they both relate to the need for and the ease of modifying the system. An example of flexibility is the situation where a system is being designed for one division of a multi-division organization. In this case it may be possible to design a system which can be used by many divisions. The system may therefore contain more facilities than required by any one division but the cost advantage to the total organization may be considerable compared to the alternative of separate but similar systems for each division. This approach can have an impact on user requirements by necessitating standardization among divisions on such matters as the chart of accounts, code design and perhaps input and output data contents so that other objectives such as efficiency can be met.

8.3.2 Schedule and Cost

In some situations the schedule or cost constraints may be of overriding importance and act as a significant factor in making design decisions. An example of a schedule constraint may be a government department entrusted with introducing a new scheme which has been promised to the public commencing on a certain date. An objective to minimize cost is frequently the case in one-off, limited life systems and leads to short cuts which would under other circumstances not be acceptable.

8.3.3 Efficiency

Efficiency for a system can be defined as the resource usage needed to run the system. Efficiency was an important objective in the 1960s and early 1970s when, motivated by high computer costs, designers would trade many hours of design time to effect marginal improvements in system efficiency. Today in many organizations efficiency is not given great weight as it is considered cheaper to utilize hardware inefficiently than to expand costly software development resources in order to tune application systems to run more efficiently. Nevertheless, in some circumstances efficiency is a very important consideration, particularly in real-time application where efficiency measures like terminal response time are critical to the success of the system.

8.3.4 Integration

When designing a system the analyst must remember that the system is a part of the total organization and is, therefore, a sub-system which must fit with the other sub-systems in that organization. Regardless of the nature of the system under study, the output of this system will provide input to some other system. Consequently, the designer must decide on the means and the extent to which the system under study will mesh with other systems. This can affect the output provided and input used, as well as the manner in which the data is stored and aggregated. Examples of increasing levels of systems integration options are:

- The same data is input to more than one system
- Printed computer output is re-coded and key entered
- Output tape file provides connection
- Master file is updated by one system and accessed by another system.
- Module level integration

8.3.5 Security

The design of a secure system involves many facets including system controls and physical security. The important point to note is that adequate control must be built into the system from the beginning, as attempts to add controls after the design is complete are usually unsuccessful and expensive. For this reason many designers welcome the assistance of audit personnel during the design stage in order to have expert advice on the level of security required and the necessary controls for the system. The need for security differs markedly from system to system. For example a cash dispensing system or electronic funds transfer system will have a far greater need for security than, say, a production control system in a cement plant.

8.3.6   Reliability

System reliability includes accuracy of the stored data, recovery and restart capability, hardware reliability, and application and system software reliability. Furthermore, increasing reliability incurs increasing costs and, therefore, decisions are necessary as to the cost/benefits of providing increased reliability in a given situation.

A common example is in a critical real-time environment, such as an on-line banking system, where system failure may interrupt customer withdrawals and so may severely disrupt and damage the organization. Because of the critical nature of the processing, hardware duplication is used to ensure close to 100 per cent hardware reliability. Increased reliability of application software is achieved through comprehensive testing by various methods and provision of means to back out transactions causing a systems crash. Also the design philosophy used needs to permit easy testing. An unnecessarily complex design will make testing more difficult and, therefore, increase the likelihood of flaws in the system.

8.3.7   Portability

Although complete portability of a system from one hardware environment to another is rarely possible, choices made throughout the system lifecycle will influence the degree of portability attained. The concern for portability arises because of uncertainty as to one's future computer supplier and the need for compatability with other computers in the organization. When designing systems, the differences between manufacturers should be recognized and borne in mind when manufacturer-unique facilities are considered in the design. Unfortunately, the use of more advanced manufacturer supplied system software, such as database management systems or transaction processing software, has reduced the degree of application systems' portability. When portability is an important objective, a standard high level programming language should be used and file access and storage strategies kept simple.

8.3.8 Simplicity

Just as some of the most effective inventions have been the simplest, so too are simple solutions to system problems the most effective. Unwarranted complexity in the design may boost the ego of the designer, but it does so at the expense of the user and the people responsible for the system's implementation. For systems designed to be operated by untrained users, simplicity of the user interface is also a very important consideration.

8.4 STUDYING THE CURRENT PHYSICAL SYSTEM

The next step in analysis after defining the system's scope and objectives is to thoroughly understand and document the current system at the physical level. An important distinction in this methodology is that of the physical and the logical system. The physical level contains complete details of not only what is happening but how it is being carried out. The logical level focuses attention only on the 'what' and omits mention of 'how'.

A simple example may help to make this point. Suppose an analyst seeks to draw a physical data flow diagram (DFD) of a manual stock recording system (refer to Chapter 3 for a discussion of DFDs). Jim B is the stock issue clerk and Jack F handles stock receipts. The stock receipts and stock issue documents are sent to Fred W who sorts and batches them before updating the stock file which is held on cards in product name sequence. The DFD is illustrated in Figure 8–3. The physical details here are:

- The names of the people responsible for activities
- The sort and batch procedures
- The sequence of the stock file

The logical DFD is obtained by removing the physical details or else replacing them with their logical equivalent. This gives rise to Figure 8–4.

In addition to documenting the physical flow of data through the system, the transaction volumes for each data input document and the number of records to be stored for each file need to be measured and recorded. Care should be taken that transaction rates are also recorded at peak times so that both the average rate and the peak rate are known. This is particularly relevant for on-line systems where there is little opportunity to smooth the load. It is also a consideration with some batch

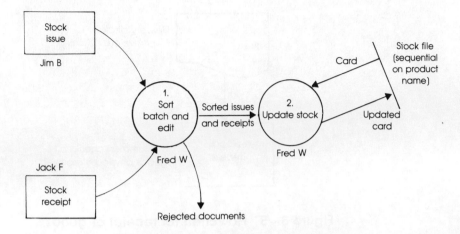

Figure 8—3   Current physical DFD

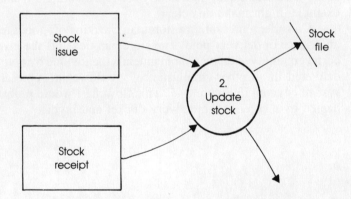

Figure 8—4   Current logical DFD

systems which exhibit strong monthly seasonal fluctuations in volume, such as the retail industry.

Obtaining the detailed data on a system, and organizing and presenting this information is a tedious and time-consuming task that calls for much patience on the part of the analyst. Part of the time spent is due to the large number of people who must be interviewed (since no one person understands the system) and the time taken in the interview. The presence in most systems of multitudes of special rules (e.g. 'this is what we do unless it is the last Friday of a month when we handle it differently') also adds to the difficulty, as do conflicting accounts of operational details.

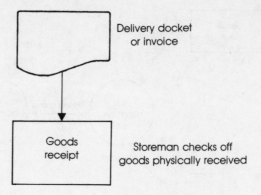

Figure 8—5   Flowchart for receipt of goods

While either a DFD or a flowchart may be used to document a system at the physical level, experience suggests a flowchart may be the more easily used technique, particularly for representing a manual system. An example might make this clear.

Consider representing a storeman checking off goods received against an invoice or delivery note. The storeman ticks off the invoice line if the correct quantity is delivered but amends the invoice by noting the quantity delivered in all other circumstances. A flowchart representing this sub-system is given in Figure 8–5. This flowchart would probably be accompanied by a copy of the delivery docket and invoice.

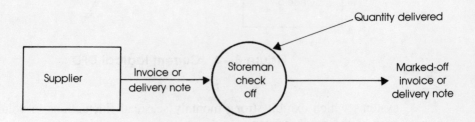

Figure 8—6   DFD for receipt of goods

The DFD for this sub-system is given in Figure 8–6. The data dictionary for the DFD is given below:

Invoice:

    Supplier name
    Address
    Invoice number
    Date
    Invoice line:*
        product code
        product description
        quantity delivered
        unit price
        extended price
    Invoice total

Delivery note:

    Supplier name
    Address
    Date
    Delivery note number
    Delivery note line:*
        product code
        product description
        quantity delivered

Quantity delivered:

    Invoice number
    Product description
    Product code
    Quantity delivered

Marked-off invoice:

    Supplier name
    Address
    Invoice number
    Date
    Marked invoice line:*
        product code
        product description
        quantity delivered
        unit price
        extended price
        quantity delivered
    Invoice total

The 'marked-off invoice' data items are written out in full to declare that the quantity received is recorded against each invoice line. A similar notation would need to be made in the 'marked-off delivery note' data flow.

The chief problem with the physical DFD is its data dictionary which becomes very lengthy as more and more needless and redundant data must be included. In the flowchart much of this information is implied or is conveyed by English text. However, in the documentation of the logical system, the DFD is the preferred choice with the data dictionary containing only the necessary logical data. This point will be discussed in the next section.

8.5 THE CURRENT LOGICAL SYSTEM

The current logical DFD is derived from the current physical system by removing all physical processing and file reference, all redundant data items, and routine processing steps like input edits. The current logical DFD shows only the necessary logical actions to transform the current input data to the current output data — no modifications of system function should be made. Hence, for example, the file contents are unaltered but the access method — being a physical attribute — is not referred to. Figure 8–4 gives an example of a logical DFD.

As a second example the current logical DFD and data dictionary of the goods receipt example of Section 8.4 is presented. The DFD is as given in Figure 8–6, but the data dictionary is much altered.

> *Invoice:* (or delivery note)
> > Invoice number
> > Supplier name
> > Invoice line:*
> > > product code
> > > quantity
> *Quantity delivered:*
> > Invoice number
> > Supplier name
> > > product code*
> > quantity received
> *Marked-off invoice:* (or marked-off delivery note)
> > Invoice number
> > Supplier name
> > Marked invoice line:*
> > > product code
> > > quantity
> > > quantity received

It has been assumed here that only the quantity information is logically needed in this sub-system. Information has been removed from the data dictionary either because it is redundant (e.g. description, address) or because it is not logically needed in this sub-system (e.g. price, extended price). This means the data dictionary for the logical system can be quite compact.

At the logical level there is generally very little difference whether a system is manual, batch, on-line real-time, conventional files or database. Note that all these descriptions refer to alternative physical processing means.

Since the current logical DFD expresses necessary actions that must be performed on the data, it forms the foundation for constructing the new system. If we are dealing with an operation-type system like order entry, purchasing or debtors, the current logical DFD will relate to standard business practice which will almost certainly need to be performed in the new system. For higher level systems like a management information system (MIS) the current logical DFD may form less of a framework for the new system.

Note that the edit, which is a logical procedure, has also been deleted in Figure 8–4. This has been omitted because all input data must always be edited and so the edit procedure contributes no additional information. At the logical level only the necessary procedures which transform the input data into output data are represented.

For simple systems it may appear unnecessary to plod through all the details of the physical system and draw a DFD or a flowchart, but experience shows that it is easy to omit important considerations (such as the impact of a new system on the users) because no one has bothered to find out in detail how the current system operates. Furthermore a detailed knowledge of the current system will be needed to plan the appropriate change strategy for implementing the new system. Diagramming the physical system forces the analyst to consider every possible aspect within the scope of the new system, and simplifies the user's job in understanding the documentation and spotting errors if they exist.

Brief notes explaining the nature of each of the DFD or flowchart procedures need to accompany the diagram. These notes should explain the actions taken to process the incoming data to each procedure and convert it to the output. In addition, copies of all the forms used in the current system need to be collected and identified with the relevant section of the flowchart or DFD.

8.6   DEFINING THE USER REQUIREMENTS

Defining the user requirements accurately and completely is probably the most essential task in the whole systems project and one most critical to its success. Small errors or omissions made in the definition of a system's requirements but only detected after coding may be impossible to rectify without a complete re-write of the system. Section 8.2 covered the important aspect of building a creative team capable of developing a sound and imaginative requirements definition. Broadly, this team should analyze the current system and the business environment spanned by it to identify the problems and opportunities that can be overcome and exploited by the new system.

8.6.1 Current Systems Problems

Deficiencies of a technical or efficiency nature are probably the easiest problems to define with regard to a current system. These may be associated with problems caused by the type of input media used or the volume of processing — thus overloading certain areas of the organization or exceeding the capacity of the current system leading to late reports and delayed processing. Also quite easily identifiable are the control problems of the current system. Auditors frequently draw these to the attention of management as a result of the financial audit.

Timing, completeness or the arrangement of the current system reports are problems frequently cited by users. New data items can be identified, preferred timing of reports specified and data re-arranged. However, the analysis team should be careful not to settle too early on these type of requirements because often they are fairly superficial and tend not to represent the major potential benefits of the new system.

8.6.2 Opportunities

The more difficult but more potentially rewarding area for user requirements lies in identifying the opportunities for a new system to make a major impact on a key area associated with the success of the business. Take the following example.

A team was involved in carrying out an investigation for a new payroll system. Initially the system's requirements were the need to carry out the same logical functions as in the current system. Minor clerical savings were identified but it was unlikely the new system would have much impact on the organization. Then the analysis team widened their scope of investigation to consider ways in which a system containing employee data might help the organization accomplish its objectives. This revealed a problem the organization had with employee record-keeping and in identifying employees with needed skills. Extending the system to meet this need represented a very significant benefit to the organization.

There is no easy recipe for spotting significant systems opportunities. The likelihood increases if members of the analysis team and its steering committee thoroughly understand the business that their organization is engaged in, its strategic plans and its current systems.

8.7 IDENTIFICATION OF DESIGN ALTERNATIVES

There are in general many solutions to a given specification of user needs, each alternative differing in some significant way from the others. Never should one solution be put forward without specifying and analyzing a number of potential answers to the system's needs. Only by doing this can the analysis team be sure that the most attractive solution has been selected. Clearly if the potential solutions proposed by the team are too narrow in their range then the solution recommended may be deficient. Thus it is important that a sufficiently broad set of solutions be considered. This can be helped by recognizing some of the main decision variables in developing alternative solutions. These include:

- The location of data input, output and processing
- The response rate of the processing
- Mode and medium of data input
- Style and medium of data output
- The system's man-computer boundary
- The level of data aggregation in the system

In addition, the technical objectives covered under Section 8.3 have a major bearing on the identification, analysis and selection of design alternatives.

Location of input, output and processing. The broad classes of solution alternatives dependent on the location of equipment were presented in Chapter 6:

1. Centralized data input, processing and output
2. Decentralized input and output with centralized processing
3. Decentralized input and centralized processing and output

For each of these alternatives there exist a number of possibilities for handling the data communications between the decentralized source of data and the central office. These alternatives include leased line telecommunications facilities, overnight mail bag, dial-up line and courier.

The response rate of processing. Processing can be instantaneous as in on-line real-time processing where each transaction is edited and its contents update the relevant files in real-time. An alternative is to carry out the editing in real-time but perform the updating on a batch processing basis. The slowest alternative is to enter and process the data in a batch mode, with a number of possible run frequencies.

Mode and medium of data input. There are a number of systems in use today which owe their existence to a clever solution to data input. An example is the on-line automated grocery checkout counter at which bar coded products are read by a laser scanner, the purchases totalled and the inventory file updated. As systems move further into the factory and warehouse, data input becomes a more difficult aspect calling for new initiatives.

Data output. Data can be output by voice, on paper, on visual display units with or without a hard copy device or microfiche, in colour or black and white, in pictures or in numbers. There are many alternatives to help the user best profit from the information produced by the system.

Man-computer boundary. Many alternative boundaries between the manual and the computer system can be drawn, each one presenting a different cost/benefit tradeoff.

Level of data aggregation. At many points in a system decisions can be made concerning the level of data aggregation. Some of these impact the reports that can be produced, others impact the future usefulness of the historical data for analysis purposes, and many impact the file size, volume of data input and the coding and data entry costs. Many retail systems operate with input and reports aggregated on a department basis as a means of reducing coding and data input costs. While there is a trend today to enter and store data on the lowest level, there are many situations where this is an expensive luxury, unjustified by the use to which the data is put.

8.8 ANALYSIS OF DESIGN ALTERNATIVES

After identifying a number of design alternatives, the next step is to narrow down the search by applying broad decision criteria, and so reduce the list to a manageable number of alternatives. What is manageable depends on the resources available for the analysis and the importance of the project. In some cases perhaps two alternatives are analyzed in detail, while in others four or five may be considered.

The general structure suggested for the analysis of each alternative is to: (1) develop a brief new logical system design, and (2) a new physical system design. There is no time to develop a complete physical design for each of the alternatives, so only the key points can be established. These key points need to be the facets of the system which have a governing influence on the way the alternative functions, and on its cost and benefits. In general, the use of a certain physical component which dominates the cost/benefit equation for the system may be the key point. An example of this is using a point-of-sale terminal to capture cheaply the product codes of goods sold. When the new physical system for each alternative has been considered (3) the technical feasibility and selection of the hardware and systems software proposed has to be argued, (4) the size and speed of the communications facility must be established, and (5) the various personnel and staffing levels need to be considered. Finally, (6) each alternative's impact on the organization and its business needs should be carefully analyzed as it is on this analysis that the system's success depends.

8.9   REVIEW QUESTIONS

1. Differentiate between top-down and bottom-up design methodologies.

2. Discuss the difference between logical and physical views of a system.

3. Outline the ways in which a user might participate in the various stages of systems development. Contrast the user's role with that of the systems analyst.

4. Why is it important to define a system's scope and objectives before any work starts on analysis?

5. Explain, using practical illustrations, the concepts of:

   (a) Flexibility
   (b) Maintainability
   (c) Efficiency
   (d) Reliability
   (e) Security
   (f) Portability

6. The De Laroux Electricity Supply Authority is considering a change to its customer billing system. The current system is as follows:

   Customer service falls into five classes:

   - Domestic–continuous
   - Domestic–off peak
   - Commercial–light continuous
   - Commercial–heavy continuous
   - Commercial–off peak

Customers may have more than one class of supply and will have a separate meter for each class. Supply is charged for at a rate per kilowatt hour, and each of the five classes has its own unique rate. The meters record cumulative usage in units of 1 kilowatt hour. The system produces customer statements in the form shown below following the periodic reading of the meter.

De Laroux Electricity Supply Authority
Statement of Account for Electricity Supply

Cust. A/c No: JON 14736B

Cust. Name: R E JONES   Period ended: 30 MAY 1983

Address: 14 VOLVO TERRACE
               MT SUPERIOR   Date payment due: 30 JUNE 1983
               DE LAROUX 9999

SUPPLY CLASS	METER	OPENING READING	CLOSING READING	UNITS CONSUMED	RATE $	CLASS CHARGE
1	74943	9875	10250	375	0.04	15.00
2	86493	586	942	356	0.03	10.68

TOTAL THIS A/C	25.68
BALANCE OUTSTANDING FROM PREVIOUS ACCOUNT	10.23
AMOUNT NOW DUE	35.91

Three files are used in the system: a customer master file, a rate structure file and an accounts receivable file.

The rate structure file comprises:

Supply class code
Rate

The accounts receivable file comprises:

Customer code
Outstanding balance over 60 days
Outstanding balance over 30 days
Outstanding balance up to 30 days

One input is the meter reading dockets containing:

Meter number
Ending meter reading
Date of meter reading

Other input to the system includes:

(a) File maintenance transactions which add, delete or modify records on the rate structure file, the accounts receivable file or the customer master file.

(b) Payment receipt transactions which are prepared and batched for updating the accounts receivable file. This consists of customer code and amount.

In addition to the customer statements, the system provides the following management report:

> A listing of all customers with amounts owing in excess of 60 days. This report lists customer code, customer name and address, amount owing over 60 days and total amount owing.

Questions

(a) Draw a set of logical data flow diagrams of the current system.

(b) List in a data dictionary the data elements that are required on the customer master file, and all data flows.

(Note the customer master file record content is not given. You will need to determine the data items necessary in each record to support the above processing.)

7. Refer to the case study in Chapter 11, Sections 11.1 to 11.5. This gives an overview of a feasibility study. Identify those aspects which would need further details before management could decide on the best course of action. State the benefits that would accrue from this additional information and consider the costs of the analyst providing the information.

9 Logical Design

9.1 INTRODUCTION

This chapter considers logical design, the next stage in the system's development lifecycle after completion of the feasibility study. If unsure of the overall systems analysis and design framework presented in Section 8.1 the reader should review this information.

The logical design document shows what the system is required to do to fulfil the user's requirements. It does not show how these functions and procedures will be implemented. In the top-down design approach these decisions are left for the physical design stage. In fact the logical design should not, in so far as is possible, constrain the physical design choices further than they have already been constrained in the feasibility study. Generally the target computer system and communications network approach are decided in the feasibility study and form the computer environment for the logical design.

While the user's requirements are largely defined in the feasibility study, the definition continues through the design stage as the user participates in the development of the design and sees ways in which the utility of the system can be increased. One part of the user's requirements is logical in nature (e.g. report contents — input, file updating), though another is physical (processing volumes and speeds, input media objectives, response times, machine size restrictions, reporting frequency and file sizes). These physical requirements are considered at the physical design stage after the logical design has been completed.

The design of the logical system needs to include all the user's logical requirements and it shows — using data flow diagrams (or another technique), the data dictionary and supporting documentation — the network of logical steps required to transform the input data to the output reports, documents and file data.

It is important that the probable general shape of the target physical environment is available to the design team before they embark on designing the new logical system, as there is a relationship between the logical system's requirements and the physical means used to accomplish these requirements. For example a user may well dictate different information requirements under different assumptions about reporting frequencies and form of presentation. The feasibility study, containing abbreviated new logical and physical designs for each solution alternative, should normally provide an adequate framework of the selected physical system for the new logical systems analysis and design to be carried out. However, this needs to be reviewed and any ambiguities resolved before commencing the logical design.

The aim in developing the logical design document should be to provide one that contains all the information necessary for the physical design team to do their job. It should, at least in theory, not be necessary to consult the user to carry out the physical design, all the required information being in the feasibility study and the logical design documentation.

The logical design of a system consists of a number of separate but related activities:

- Design files
- Develop design
- Plan implementation
- Design controls
- Document the logical design
- Develop data dictionary
- Walkthrough the design

9.2 DESIGNING FILES

Designing the files for the new logical system requires grouping into like classes the objects or entities for which the system needs to maintain or store information. Examples of entities typically encountered in business systems are customers, products, employees, sales orders, machines, back orders, assets and general ledger accounts. Some of these entities relate to groups of transactions and would be stored on a transaction file, while others relate to entities which are relatively long lived (e.g. customers, products) and would be stored on master files. In this chapter it is assumed for simplicity that the data structures are not complex.

In most cases the individual records need to be identified either individually (e.g. master file records) or as a member of a group (e.g. transaction file records). Examples of group membership are:

- All the order transactions in a certain batch or received in a certain period of time
- All the orders for a certain product
- All the orders from a certain customer for a certain product

A record which needs to be accessed individually is identified by the unique record key which is composed of one or more data items on the record. For example a customer number is normally the data item which acts as the key for the customer file. As the key constitutes the unique identifier for a record it should be carefully designed. A key may be chosen from one of three classes: structured, unstructured or partially structured.

Structured or speaking keys have certain positions within the key structure designed to communicate coded information. For example, 1250N may indicate a 50mm long, 2mm diameter nail. Here the second digit gives the gauge, the next two the length in millimetres and the final character indicates we are dealing with a nail. These keys were popular with unit-record-oriented systems when much sorting was necessary because only sequential files were available, and record storage was very much at a premium (record length was generally preferred to be shorter than the 80 characters of a punch card). Another example of a structured key is the airline three-alphabetical-character airport code. Thus SYD is Sydney, SFO is San Francisco, LON is London and NYK is New York Kennedy Airport.

Unstructured keys contain no information apart from the fact that they refer to a certain object. When unstructured keys are used all the data concerning the object are kept as attribute information within the data items of the record. Hence, if the nails just referred to were given an unstructured key the length, gauge and product type would be held as data items within the record.

Partially structured keys contain some unstructured information. For example a student identification number may hold the year of first registration in the leading two fields with the remaining fields unstructured.

Desirable properties of keys are: they should be brief, not subject to change, unique and simple.

Brevity. Keys should always be as short as possible. However, their length must be great enough to uniquely identify all the objects, and allow for growth and turnover within the objects. This last point refers to the situation where entity records are being deleted and added but the deleted keys cannot be reused for some years. If keys are longer than necessary, extra clerical time is taken transcribing and entering the key data, additional errors are made, and storage space is needlessly occupied.

Unchangeability. This characteristic applies principally to structured keys which should not hold data that are likely to change over time since it is a major undertaking to change a key for all the history, budgets and other related data. An example of a poorly designed code is a product code with a leading field which indicates the selling division responsible for that product. If, at a later stage, the organization wishes to re-organize the selling responsibility for its products, its coding structure would be a barrier.

Uniqueness is an obvious requirement for a key that is to provide an identification of a single object.

Simplicity can relate to both the key structure and its length. Keys should desirably all have the same number of fields to avoid the problems of right or left justification and of recognition when a character of the key has been unintentionally omitted. It makes for a simpler key structure to have all fields of either numerical or alphabetical characters. Mixed alpha and numerics cause confusion between such pairs as 7 and T, 2 and Z, 6 and G, 5 and S and a number of others. One has only to try to decipher a hand-written British postcode to see the problem.

Example

As an example of designing logical files, consider the task for the following data items, all related to a customer:

> Customer name
> Customer address
> Customer payment amount and date
> Customer industry code
> Customer industry name
> Purchase order:
>> date
>> product code
>> product description
>> quantity
>> cost
> Credit rating

The first design task is to identify the entities. The first entity is clearly a customer. All the descriptive data or attributes for a customer will be collected together, as listed below and called 'customer'.

> *Customer:*
> Customer name
> Customer address
> Customer industry code
> Customer industry name
> Credit rating

Now it is assumed that many customers can belong to the one industry code and that each industry code has the same name. Hence it is both inefficient and potentially problematic to have industry code and name in the customer record; inefficient because in each customer record the redundant industry name is stored, and problematic because if a customer changes industry code someone may forget to change both the code and name. Thus the industry name should be removed and stored in a new entity called 'industry'.

> *Industry:*
> Industry code
> Industry name

Any additional data on the industry could also be stored in this entity.

One more step needs to be taken for 'customer' — a code needs to be defined. Sufficient has been said on this code design already, so the final customer entity is written:

> *Customer:*
> Customer code
> Customer name
> Customer address
> Customer industry code
> Credit rating

Looking at the original list of data two kinds of transaction data can be recognized: payment data and purchase data. Because these are two discrete objects of interest they should be stored in separate entities. The 'payment' entity is rather obvious with the customer code inserted before the payment and date data:

> *Payment:*
> Customer code
> Customer payment amount
> Date

The 'purchase order' looks equally straightforward but it has a catch. Can you see it?

> *Purchase order:*
> Customer code
> Date
> Product code
> Product description
> Quantity
> Cost

The product description is redundant data as product code suffices logically to identify the product. Thus the description can be removed. However, in the event that the 'purchase order' entity was an output field written to a purchase form, the description would obviously be needed. But for storing the transaction data it can logically be dropped.

9.3 DEVELOP THE DESIGN

For most data-processing systems, business practice dictates that the current logical system forms the basis for the new system. For example the steps that must be carried out to process an order, adjust stock levels and produce an invoice would reflect the traditions, policies and standard practices of a particular firm and industry, and are largely independent of the processing technology used. Thus in designing a new system the starting point is the documentation of the current logical system — only rarely is there no current system.

The current logical system generally forms the framework or skeleton into which the user's new requirements are merged. While preserving the logical structure of the current system the design effort frequently requires substantial modifications. These may involve modified or new inputs, data flows, files and outputs, and the insertion of new or modified processes. This results in extension and modification of current DFDs, but in most cases preserves the broad form of the diagram. For an example of developing a new logical DFD, consider the current logical DFD for the elementary (manual) order entry system shown by solid lines in Figure 9–1. It is assumed the new requirements can be summarized as:

- Place this system on the computer to enable the processing of increased transaction volumes
- Produce listings of orders and back orders

These new requirements, shown in Figure 9–1 by dotted lines, produce only minor modifications to the current logical system. They also require additional data to be captured on the files, and a number of modifications to the data dictionary and to the detailed procedure specifications. The case study in Chapter 11 presents a more complete practical description of the design process.

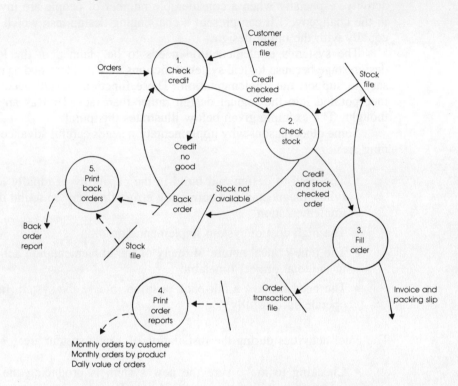

Figure 9—1   Order entry system logical DFD

9.4   PLAN IMPLEMENTATION

A careful and well-devised plan for the implementation of a new system needs to be made at the logical design stage. An important reason is that the system should be designed so that the changes can be introduced to the user in a way that is most likely to lead to the system's acceptance. A bad impression created during the system's implementation by poor training, excessive errors, long breakdowns or any other reason, may jeopardize user support and cooperation and require much work to over-come. Any system introduces change to its users, and research shows that the way the change is introduced is as important as the change itself in influencing attitudes to the system. Bostrom and Heinen (1977) give a good summary of this research. The process of introducing change starts at the project initiation and reaches a climax during implementation, when the users gain their first impression of the operational system. Planning a smooth introduction for a new system is often a major and complex activity, especially when a considerable number of people are involved in the changeover. It can present a challenging design task which ranks equally with the initial design.

The system's implementation needs to be planned in the logical design stage because logical system functions may be identified as necessary to support the implementation. These functions would need to be incorporated into the logical design rather than tacked on as an afterthought. The example given below illustrates this point.

Some other reasons why implementation needs careful advance planning are:

- The need to design and build-in the capability of rapidly identifying, diagnosing and correcting errors that may be found during implementation
- The high cost of system implementation
- The time-critical nature of many of the implementation activities in the total project timetable
- The need to have a fall-back position in case the system fails to operate successfully

The chief activities during the installation of a new system are:

- Checking to make sure the new system is producing the right information within the specified time frame
- Completing the training of users, key entry operators and computer operators
- Training management in the use of the information reports produced by the system

To illustrate the need for early implementation planning consider an example drawn from a case study developed by one of the authors. A large international oil company developed a new capital assets system to give better control over its widely distributed assets of service stations, bulk depots, refineries, offices, plant and equipment. The new system, an advanced database application, was to replace a simple sequential file application which had served the corporation for more than ten years. No thought was given to the implementation of the new system during the design phase, as it was believed that parallel running of the two systems would be straightforward. The unexpected problem which occurred in the implementation was the extreme difficulty of reconciling the two systems. Since all the reports had been changed it was very difficult to trace imbalances in totals, except by laborious manual searches through file dumps. Some 30 clerks were co-opted for three weeks each month to reconcile the two systems' monthly reports in order to prove the new system correct. After three months of parallel running, the chief accountant could not stand the strain on clerical resources any longer and, despite the fact that a number of errors were still being located, decided that the new system had to be satisfactory. In fact quite a number of errors located were due to problems in the old system, not the new system. Had thought been given to the task of reconciliation, special computer-prepared reconciliation reports could have been provided for in the system and the manual effort almost entirely eliminated.

9.4.1 Changeover Approaches

The changeover approaches which generally form the basis for implementation plans are:

- Parallel running of new and old systems
- Phase in new system and phase out old system
- Pilot study of new system
- Start up new system and stop old system

These four methods are illustrated in Figure 9–2.

Parallel run changeover. In this approach both the new and the old system are in parallel for a number of cycles until the new system has been proved to be operating reliably and correctly. This changeover approach is generally suitable where the new system is not very labour-intensive so that operating both systems together does not over-tax employee resources.

However, it is inevitable that parallel running will involve a number of staff in operating two systems concurrently, and the implementation plan must provide for temporary staff and/or overtime to catch up on the extra work load. Because parallel running involves starting the whole new system up at one time, it is only suitable for small to medium-sized systems. For large systems, sub-systems need to be implemented progressively both because of the effort required to get a system going and also to localize problems to a smaller more manageable segment.

Phased changeover. Phased changeover is the process of progressively cutting small segments of the old system across to the new. This may be done by periods of parallel running for each segment before cutting across or, alternatively, cutting a segment across directly without running in parallel first. Deciding whether to run in parallel depends on

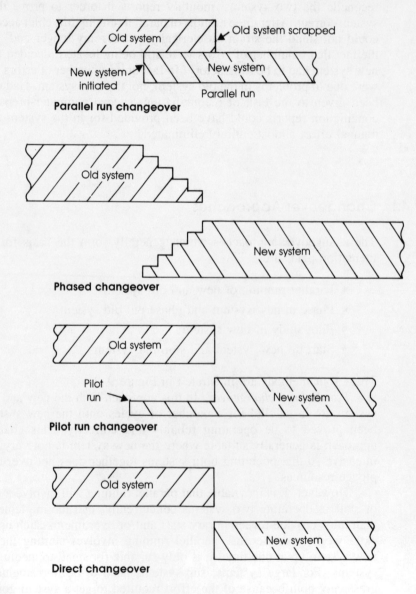

Parallel run changeover

Phased changeover

Pilot run changeover

Direct changeover

Figure 9—2 Changeover approaches

the scale of the segment, the difficulty of returning to the old system should the new be unsatisfactory and the impact on the organization of the system being down for a time.

This changeover approach is most suitable for large projects and for projects involving a significant number of personnel, when it is impractical to consider operating the old and new systems in parallel. The phased changeover is particularly suitable for installing a system where there are natural segments in the user population such as provided by an organization structure. For example, when installing a new payroll system for an organization based on a divisional structure, it may be decided to phase the system installation by division.

Pilot run changeover. The pilot run method uses the new system to rerun data from a previous period, thus allowing reconciliation of the data. Alternatively, sampling a selection of the current data may be used as input to the system.

As the name implies, this method is in essence a further system test on live data. When the system has been proved to be working correctly and the staff are comfortable with operating it, the new system can be started up and the old stopped.

Direct changeover. The direct changeover method is to stop the old system on (say) the last day of the week, month or year, and start the new system up. In some circumstances there is no alternative to this method. However, it must be borne in mind that direct changeover is risky unless the old system can be easily revived and used as a fallback position. This is often impossible as the old system's files are out of date. Generally, if this method is required, a period of pilot running should be considered to first test and prove the system under operational conditions.

A number of companies in Australia have been in very great difficulty as a result of using direct changeover for their accounts receivable system. Faced with the new system not working, they were unable to send statements to their customers for a period of some months, thus placing the whole organization at risk.

9.5   DESIGN CONTROLS

Operational controls, that is the input, processing, file and output controls for the system, need to be designed and documented. In the logical design their documentation consists of recording the relevant control processes and flows in the DFD and incorporating the details of the flows and output data items in the data dictionary. See Chapter 7 for a discussion of the principles of designing operational controls.

9.6   DOCUMENT LOGICAL DESIGN

The documentation of the logical design consists principally of the DFDs, the data dictionary, descriptions in natural language or in flowcharts of the processes, and associated written comments to explain and communicate the design. The documentation techniques of DFDs and data dictionary allow the design to be incrementally recorded as it is developed, so apart from some 'prettying up', the final documentation of the design should be available immediately on completion of the task.

The final data dictionary contains entries for all input documents, output forms (including control outputs), files and data flows between processes of the DFD. In some cases management may want roughed-out input and output forms in order to appreciate the legibility of the information, and these can be prepared.

9.7   WALKTHROUGH THE DESIGN

A walkthrough is conducted by peers associated with the project carefully reading and logically checking the design by stepping through, in this case, the DFD and making sure that all the required data are present and that data are not created from nothing or do not disappear without trace.

One of the important activities in checking the new logical DFD is making sure that the diagram is internally consistent. That is, for each bubble (circle) in the diagram, all the data that comes out also goes in. Consider the exmple ion Figure 9–3 (refer to Section 7.2.3 for definition of b/f and c/f). In the dictionary the data flows are defined as:

> Purchase invoice details:
>> product code
>> date
>> quantity received
>> total cost
>
> Old stock details:
>> product code
>> b/f quantity on hand
>> b/f average cost
>
> New stock details:
>> product code
>> c/f quantity on hand
>> c/f average cost

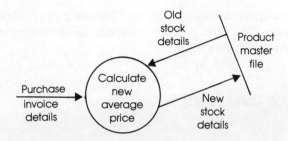

Figure 9—3   Consistency check sample

The documentation for the bubble 'calculate new average price' shows that:

$$\text{c/f average cost} = \frac{(\text{b/f average cost} \times \text{b/f quantity on hand}) + \text{total cost}}{\text{c/f quantity on hand}}$$

$$\text{c/f quantity on hand} = \text{b/f quantity on hand} + \text{quantity received}$$

A check such as is shown in Table 9–1 can prove the bubble flows consistent by demonstrating that each of the output data items can be traced, either directly or through a calculation, to input data. Such a check is needed for all the non-trivial bubbles in the DFDs and for all flows to and from files. This helps prevent module and program interface problems which normally contribute so significantly to errors found in program and system testing.

Output data items	Input data items used (if a calculated field)	Source
New stock details:		
product code		purchase invoice details
c/f quantity on hand	b/f quantity on hand	old stock details
	quantity received	purchase invoice details
c/f average cost	b/f average cost	old stock details
	b/f quantity on hand	old stock details
	total cost	purchase invoice details
	c/f quantity on hand	source already specified

Table 9—1   Consistency check of bubble 'calculate new average price'

Data integrity controls for input, files and outputs should be specified in this stage. The principles for designing these features are discussed in the data controls sections.

9.8 REVIEW QUESTIONS

1. A professional society with branches in each State has a national membership system. The key for each member is an eight digit integer, where the leading two digits identify the member's state, and the last digit is a modulus 11 check digit. Comment on this key structure.

2. The following data items are all related to students:

> Student registration number
> Student name
> Student address
> Course code (major)
> Subjects currently enrolled*
> Student history:*
>> subject number
>> grade
>> year
> Textbooks for each subject*

* Indicates repeating item or data structure: see Chapter 4.

Some of the data entries above are descriptive but not sufficiently precise to designate the data items. Design the logical files necessary for this data, being specific about the data items in each file. Note that a requirement of this design is that all records are of fixed length.

3. Modify the design shown in Figure 9–1 to include the requirement that orders be processed once a day and that priority for stock be given to customers according to their category code (A,B,C or D) held on the customer master file.

4. Give examples of systems and environments where each of the changeover approaches listed below would be preferred. Explain why it is the preferred approach in the environment outlined:

(a) Parallel run

(b) Phased changeover

(c) Pilot run

(d) Direct changeover

5. Walkthrough the logical design given in Chapter 11, Section 11.6.

   (a) Comment on any inconsistencies in the system.

   (b) Comment on the adequacy of the system controls.

   (c) Draw a data flow diagram for that portion of the system carried out by the computer (see Figures 11–10 to 11–13).

   (d) Refer to Figure 11–11: to be successful Processes 3 and 4 assume the flexibility and intelligence of a person. What additional files, processes and data flows would be necessary if these functions were to be performed by a computer?

10 Physical Design

10.1 INTRODUCTION

The physical design stage of software development aims to provide the details so that the new logical system can be implemented in the target environment. This task is usually carried out by the same team of designers who have been developing the logical design, and it occurs once approval of that logical design has been given.

While physical design activities are normally carried out after the completion of logical design, an exception occurs when some physical characteristics of the system need to be established in order to determine the feasibility of a particular design approach. An example of this might be where one design option considered in the feasibility study stage for an organization involved the use of optical character scanners that had previously never been used in the country. To establish that the new technology would work in the particular application, a company representative went to the country of manufacture and observed the equipment in operation.

The activities of the physical design stage can be divided into:

- Input/output media selection
- Developing the physical system
- Defining the physical files
- Packaging the processes
- Documenting the design

149

10.2   INPUT/OUTPUT MEDIA SELECTION

The choice of input and output media can have far-reaching cost implications for the system designed. Frequently the costs of input data capture
and output and dispersal are a significant proportion of the total operating
cost. In some cases the specialized and efficient nature of the data capture
system is the key to the cost effectiveness of the entire system. Some
point-of-sale systems are good examples of this. Other reasons for taking
care in designing the input and output systems are:

- Poor input design leads to input errors and user dissatisfaction
  with the system
- The variety of input devices available provides many alternatives,
  one of which may afford a very creative solution to input problems

10.2.1   Input Media

Input media is discussed under batched, on-line and special purpose input.

Batched input

Batched input is one of the most commonly used methods. It affords good
opportunity to verify and control input, though it isolates the user from
data preparation and introduces time delays between data entry, data processing and output. This makes error correction more difficult and reduces
motivation for careful data entry. The main methods used for batched
input can be divided into keyed input or direct reading devices.

Keyed input is still the most commonly used method, and can produce
punched cards, magnetic tape or magnetic disk. The use of punched cards
is now rare.

Key entry on to tape or disk has many advantages over punched
card. These include factors in the physical environment such as silent
operation, faster computer reading of the medium, simpler storage and
handling of the medium, and faster keying rates because of the elimination
of card transport time in the card punch. Other advantages are due to the
inclusion of buffer memory and logic capability in the key device. This
means that simple edit checks and batch controls can be carried out while
data entry is in progress. Furthermore, because the whole record is normally carried in buffer memory until it is completed to the operator's
satisfaction, if the wrong key is depressed (80 per cent of errors fall in
this category) the error can be corrected immediately.

For high-volume data-entry applications, minicomputer-based systems are often used. They enable many (say from 5 to 50) operators to key data which is then stored on a central disk unit. Batches are written to the magnetic file for input to the main computer. This is illustrated in Figure 10–1. The main advantages, in addition to those already discussed, are:

- A high degree of keying format support
- Extensive edit checking of input data
- Statistics and status reports prepared on-line for supervisor
- Operators do not have to handle diskettes or tape
- Cheaper system per key station
- Verification control can be simplified
- Automatic batch control checks

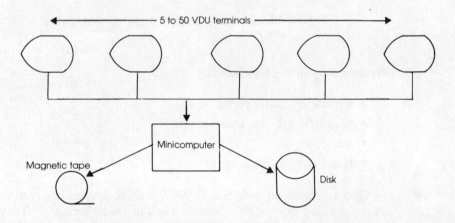

Figure 10–1   High-volume data-entry system

The main disadvantage is:

- All key stations become inoperable if the controlling minicomputer goes down

Direct-reading devices are able to decrease the time taken to input data to the system, and reduce the number of errors occurring in the data input activity, and consequently the use of these devices is increasing.

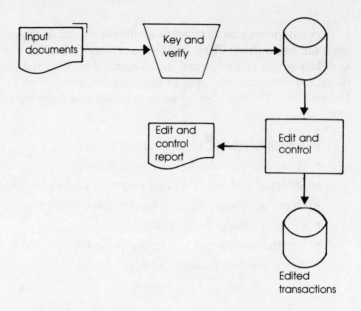

Figure 10–2 Fully keyed input procedures

Direct-reading devices include:

- Optical character recognition
- Magnetic ink character recognition
- Bar code
- Hand-marked mark sense

Optical character recognition (OCR). OCR devices can read mechanically formed or precisely written hand-printed characters. They are therefore well suited to turnaround documents such as utility accounts, having the advantage of being both computer and human readable. A major problem, however, is read errors which give rise to increased system activities. This can be seen by comparing Figure 10–2 with Figure 10–3. If fully keyed input is used only one procedure applies to the data entry task, that of keying and verifying the transactions. But if OCR input is used there is still a need for keyed input when input documents cannot be correctly read or are not available. Thus the automated and manual procedures are both necessary, adding to the system's complexity.

Magnetic ink character recognition (MICR). This technique is used mainly in the banking industry. The MICR characters are preprinted and provide low error rates, but at an increased cost.

Bar code. This code is read by a document reader, and can be used as an alternative to OCR characters on turnaround documents. The chief disadvantage is their lack of human readability.

Hand-marked mark sense. These devices operate on the basis of sensing a mark in a specific location on the document, and as such can be used for many forms of coded data. They cannot be used for free form data.

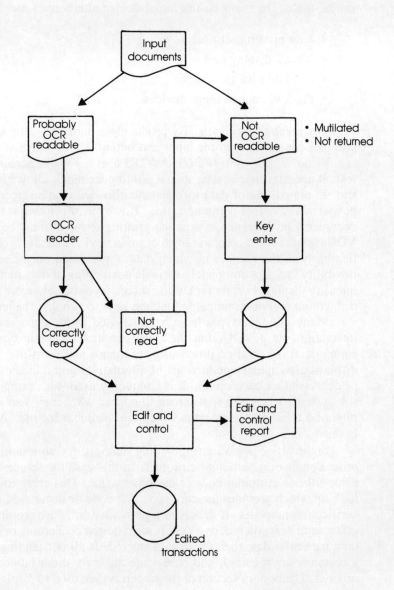

Figure 10—3   OCR input procedures

On-line input

On-line input means the use of terminal devices which input data to a computer system and allow the update of files or the editing of transaction records which are used for updating later. The big advantages of on-line data entry are the ability to correct data at time of input, and update files in real-time. The main on-line input device alternatives are:

- Key printing terminals
- Visual display unit
- Card and badge readers
- Process control input devices

Key printing terminals. By producing a hard copy, the key printing terminal can serve both for input and output.

Visual display unit (VDU). A VDU uses a TV-like screen to display text. It operates much faster than a printing terminal, allows for selective and simple editing of data formats and allows a whole screen of data to be assembled before communication. However, if printing is required for documents or reference, a separate printing terminal must be employed. VDUs are the most common form of input device and allow considerable flexibility in the manner in which data is presented on the screen. This flexibility includes completely variable positioning of data items, variable intensity display, buffers for storing screen layouts, automatic cursor control, colour and programmable function keys to simplify the terminal use.

Many dialogue types have been devised to facilitate man-machine interaction via a VDU, but the most common dialogue types used in commercial information processing are menus and form-filling. A menu displays a computer-operated list of alternatives and indicates the key to be depressed for each option. It is commonly used, for example, to select sub-systems where the first screen to appear when a system is activated might be a list of sub-systems which are available for use. An example is shown in Figure 10–4.

Figure 10–5 shows a form-filling dialogue for an inquiry about the order details of a particular customer. In this case the screen user would enter either a customer code or customer name. This entry triggers a file look-up which provides the corresponding code or name, and address for verification purposes. If details of a particular order are required then an order number is entered, or if details are required concerning orders placed on a particular date then the date is entered. If all outstanding orders for a customer are required, null entries are made for order number and date ordered. The bottom section of the screen is then used to display the order details and complete the dialogue.

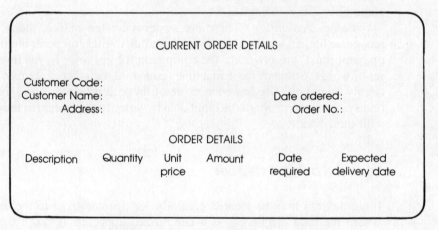

THE U ORDER SYSTEM

Code	Function
1	Data entry
2	Report production
3	File maintenance
4	Inquiries
5	File update
6	File back-up
7	Exit
SELECT:	

Figure 10—4   Main menu

CURRENT ORDER DETAILS

Customer Code:
Customer Name: Date ordered:
        Address: Order No.:

ORDER DETAILS

Description	Quantity	Unit price	Amount	Date required	Expected delivery date

Figure 10—5   Form-filling dialogue

Special purpose input

This category of input devices is custom-designed to suit the needs of particular classes of applications. They may store the data locally for later transmission to the computer, or they may be on-line and transmit direct.

There has been a significant growth in special purpose input devices as organisations seek to cheaply capture data at the point where it originates. Frequently the development of this kind of input device may open up a whole new class of application system as in the case of supermarket point-of-sale scanners.

The major categories of devices are as follows.

Card and badge readers. These are typical of special purpose terminals for point-of-sale, library systems, time-clock systems, etc.

Process control input devices. In some manufacturing applications the machines can be instrumented and attached to an input device to automatically input such data as production quantity.

Tags. Some merchandising systems use machine readable tags, either OCR or bar coded, attached to the merchandise. The tag is separated at sale time and input to a special purpose reader for recording generally on magnetic tape attached to the cash register or input directly to the computer.

Bar code or OCR wand scanners. These input devices look like and are as portable as small tape cassette machines and may also allow input from a small numerical key pad. They are finding a number of applications in merchandising, including readers for point-of-sale terminals, product replenishment for supermarkets and library book recording systems, to name only a few examples.

Voice recognition. There are systems on the market today which recognize human words and convert them into digital computer input. Each operator must 'pre-program' the equipment for his voice by reading in the set of words. Some of these machines can store the voice patterns of up to twenty operators. Use has been made of these in production and distribution systems where operators input data by voice to the system as they work with their hands.

Location of input devices

Input devices may be located centrally, or decentralized to the user site where the data originates. If a centralized approach is taken then it is posible to achieve higher data capture speeds and tighter control. However, decentralizing data capture reinforces the user's responsibility for the provision of data and, by providing an environment in which the user sees more of the computer system, may motivate him to provide more accurate data input.

Each of these approaches offers significant advantages and are widely employed in industry and commerce.

10.2.2 Output Media

The primary means of output available to the system designer are printers, VDUs, voice or microfilm/microfiche. With the exception of voice, these are all used extensively in current systems.

Printers can be classified in four ways:

- Impact or non-impact
- Dot matrix or shaped character
- Line or character
- Speed

With respect to speed, a wide range is available:

- Low speed — 10 characters per second
- Medium speed — up to 200 characters per second
- High speed — 200 to 2000 lines per minute
- Very high speed — 4000 to 18 000 lines per minute

The wide variation in speed arises because of the different technologies used by different printers. Impact line printers usually work by filling a buffer with an entire row of characters before printing the line, using a chain or drum mechanism. A chain-type line printer provides fair quality print at maximum speeds of around 1000 to 2500 lines per minute. A character printer, as the name suggests, prints one character at a time across the page. It may use a ball, a cylinder or a daisy wheel mechanism to strike the page. Another form of impact printing uses a dot matrix of five by seven or seven by nine pins to form the characters. Non-impact printers use ink-jet or xerographic principles. Ink-jet printers spray charged ink particles through an electrostatic field to direct them onto the paper; xerographic printers image the characters onto a printing surface which is then toned, i.e. ink particles are attached and then transferred to paper. Both of these methods provide very fast printing speeds. For example an ink-jet could provide up to 45 000 lines per minute using multiple jets.

Voice output is largely used in special applications such as airport flight announcing and telephone directory information. Computer output to microfilm/microfiche (COM) is widely used for price lists, parts lists, libraries, ledger account balances, etc. because it provides very high output speeds with compact information delivery.

10.3   DEVELOPING THE PHYSICAL SYSTEM

Having developed the logical system it is necessary to detail the operation of that logic on the selected hardware and in the organizational environment. This requires the complete specification of input, output, files, and processing methods and requirements. The result of this activity is the physical design specification which provides the basis for system implementation.

10.3.1 Input

The logical design provides in draft formats for the input documents and screen layouts. These formats must be converted to their final form which can be used (a) to set up the document printing in the case of hard copy, or (b) as part of the specifications for programming in the case of screen layouts. If a document is being designed, factors such as exact character positioning, type font, paper quality, paper colour and wording of instructions must be resolved. Some suggested guidelines for document design are:

- Each form must carry its own clear identification, in particular a title, name of the organization, a form number and any necessary controls (such as serial numbering, coloured multiparts, etc.).
- If instructions on filling out the form are necessary, they should be positioned so that they do not interfere with the body of the form, yet be clearly identified so that the user can see them easily.
- The form should be designed to facilitate the entry of data. For example if data are to be entered by hand, sufficient space must be allowed for a person to write them in, or if a form is to be filled in by typewriter, the number of starting positions should be kept to a minimum.
- The form should facilitate effective use of the data. In many forms horizontal or vertical lines or other separations make it difficult to recognize the response or data area which relates to the pre-printed question or description of the data item.
- The arrangement of the form should be simple. Related data should be grouped together, and non-related data shown appropriately. Quite often a form consists of three basic parts: the introductory data, the body of the form, and the conclusion such as signatures.
- Major data items should be highlighted and easy to find.
- If multiple copy documents are necessary, the identification of each copy should be obvious, which can be effected using colours or obvious copy headings.
- If the form is to be stored, then the form identification should be placed to facilitate its retrieval from the file mechanism.
- All fixed data should be preprinted on the form.

Similarly with screen layouts, factors such as exact positioning and method of display must be decided in accordance with the terminal and system software characteristics of the target hardware environment.

10.3.2 Output

Here again, the draft outputs must be converted to their final form, be they computer listings, screen layouts or partially preprinted documents. For computer listings, printer layout sheets showing the exact positioning of headings and sample entries, provide the necessary specifications for programming, while in other cases the requirements are the same as discussed for input. Details of printer and screen layouts are given below.

10.4 DEFINING THE PHYSICAL FILES

A proposed set of files with the record key(s) and contents will have been documented as a result of the logical design. They now have to be (1) confirmed or altered in accordance with the capabilities of the target computer environment, and (2) fully documented. The final decision on a file's organization, content and access paths must be made in the light of the system's processing requirements — these must therefore be considered in parallel. Leaving this aside for the moment, the system software capabilities play a major part in determining the physical form of the files. Thus IMS or DL1 database segments and almost exclusive use of VSAM files might typically be found in an IBM environment. This is not to say that organizations do not move outside the provided file management software, though the cost of doing so may not be easily justified because of the increased risk of failure and the increased expertise required by the designers and (most importantly) those responsible for implementation. See Chapter 5 for a detailed discussion on physical files.

Once the physical organization, content and access paths have been determined, file documentation should be established showing the fully-defined record contents in terms of the fields and their pictures, key field, volumes and organization.

10.5 PACKAGING THE PROCESSES

This stage of the design involves packaging the bubble procedures into programs, and specifying the systems security, control and back-up provisions in the light of the chosen hardware, file organization, input and output. Some of the broad principles which can be used to guide the program packaging (run subdivision) activity are:

- Keep programs small and homogeneous in function and data, as this should reduce programming time and later maintenance effort

- Provide for the completion of all editing before any updates are begun
- Be more concerned with the ease of subsequent maintenance than with run-time efficiency

10.6 DOCUMENTING THE DESIGN

Documentation methods of the systems analysis and design task are often also tools which assist in the various activities. Thus documenting the existing system is a task of the analysis stage and the form of the documentation will depend on the tools and techniques selected for use by the analyst.

However, any system developed needs further documentation which provides a record of the characteristics of that system and provides the basis for the system implementation. Another purpose of this documentation is to act as a vehicle of communication for the system users so that they can gain an understanding of their interface with that system. Once a system has been implemented it is subject to continuing maintenance and enhancement and requires continuing accurate documentation. This section illustrates:

- Report layouts
- Screen layouts
- Program specifications
- System specifications
- User instructions

10.6.1 Report Layouts

The contents of all reports are determined during the logical design. It is now necessary to prepare an exact specification of the layout of all reports produced by the system. The use of a 'printer layout worksheet' marked with (say) 136 columns and 67 rows enables the report format to be precisely defined and forms a part of the program specifications. Figure 10–6 is an example of such a worksheet. Once the system is operational it is advantageous to compile a folio containing both descriptions and computer-produced examples of all reports in the system, such as illustrated in Figure 10–7.

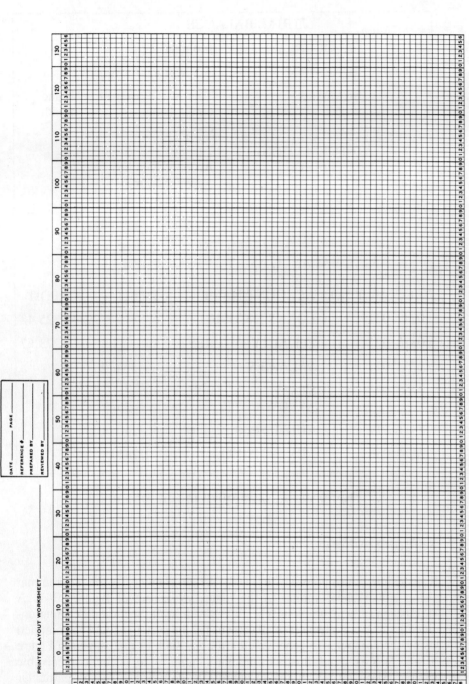

Figure 10—6 Printer layout worksheet

```
┌─────────────────────────────────────────────────────────────┐
│                      TRIAL BALANCE                            │
│        FREQUENCY:          Monthly                            │
│        FUNCTION:           Accounting Aid                     │
│        SUMMARY:            This is a register of debit and    │
│                            credit balances as at month       │
│                            end.                              │
│                                                              │
│        INFORMATION                                           │
│        SUPPLIED:           Account Number                     │
│                            Account Name                       │
│                            Balance Amount                     │
└─────────────────────────────────────────────────────────────┘
```

```
┌─────────────────────────────────────────────────────────────┐
│                      XYZ LIMITED                             │
│           TRIAL BALANCE AS AT 24/10/82      Page 1          │
│                                                              │
│  ACCOUNT NO.  ACCOUNT NAME          DEBIT        CREDIT      │
│  0347–021     LOANS INCOME                      45 325.42    │
│  0373–025     PENALTY INTEREST                     273.67    │
│  1764–021     SALARIES            17 364.78                  │
│  1768–023     STAFF                                          │
│               ALLOWANCES             435.56                  │
│  2317–314     AUDIT FEES             240.00                  │
│  2474–056     ADVERTISING          1 745.00                  │
└─────────────────────────────────────────────────────────────┘
```

Figure 10—7   Illustrative report folio entry

10.6.2 Screen Layouts

The designer must precisely define the screen contents for the programmer using a worksheet similar to the report worksheet of Figure 10–7. A discussion of screen-layout types was given in the section describing on-line input. A complete set of all screens for the system should be compiled using the computer to generate a hard copy such as that shown in Figure 10–8. This example illustrates the main menu for an accounts-receivable system, showing which key to depress to gain access to any particular part of the system.

PRESS RESPECTIVE NUMBER FOR REQUIRED FUNCTION

1   INQUIRY BY CONTRACT NUMBER
2   NEW ACCOUNTS
3   AMENDMENTS
4   CLOSE ACCOUNTS
5   POSTING
6   DAILY BALANCE SUMMARY

Figure 10—8   Menu screen

10.6.3 Program specifications

To design and code each program, the programmer needs a description of the program. This should contain: program number, program name, overview description, system flowchart, files required, input and output layouts where relevant and processing requirements. An example of a program description is given in Figure 10–9. In addition to this form of program documentation, a complete list of all programs in the system should be compiled, showing information such as program number, program description, screen numbers used and report numbers generated. This list provides cross references between input, program and output, and makes it easy to determine quickly the function of each program in the system.

10.6.4 System Specifications

The system specification document outlines the whole system at a broad level, and consists of a logical data flow diagram, system flowcharts, input documents, output documents, general description, controls and security, and implementation plans.

FIXED ASSET SYSTEM OCTOBER 1983

PROGRAM NO: FAS20

PROGRAM NAME: FIXED ASSET AMENDMENTS

PROGRAM OVERVIEW: FAS20 reads the valid Fixed Asset Transaction File created by FAS10 and updates the Fixed Asset Master File. The accounting entries generated are output to the General Ledger Transaction File and all additions, deletions and adjustments to the Master File are listed on the Fixed Asset Amendment Report.

FILES

1. Input Files
    (i) Valid Fixed Asset Transaction File — FAT6
        —Refer File Specification, page 17

2. Input-Output Files
    (i) Fixed Asset Master File — FAM2
        Refer File Specification, page 14

3. Output Files
    (i) General Ledger Transaction File — FAO8
        —Refer File Specification, page 16

REPORTS

(Here would follow layouts of the reports produced by this program.)

PROCESSING REQUIREMENTS

* When S Switch = '0' process all records
* When S Switch = '1' process only Sydney
* When S Switch = '2' process only Melbourne

Figure 10—9   Program description

Note: This is only an example; normally the processing requirements would be more detailed.

10.6.5   User Instructions

It is vital to have a set of computer-use instructions wherever a user makes contact with the system, as in decentralized data entry. These instructions assist user understanding and ensure the smooth operation of the system. Once again, this type of documentation is best explained through an example; in this case an on-line accounts-receivable system. Figure 10–10 shows the first panel displayed — the main menu.

To proceed with an inquiry relating to an account, the instruction displayed on the screen is followed, i.e. PRESS RESPECTIVE NUMBER FOR REQUIRED FUNCTION. When the ENTER key is pressed, the result is PROCEED TO INQUIRY MENU SCREEN 2 as shown in Figure 10–11.

```
SCREEN 1                    MAIN MENU              COMPANY XYZ

                PRESS RESPECTIVE KEY FOR REQUIRED FUNCTION
                      INQUIRY BY CONTRACT NUMBER
        1    – NEW ACCOUNTS/ADDITIONS
        2    – AMENDMENTS
        3    – CLOSE ACCOUNTS/DELETIONS
```

Figure 10–10   First panel display

```
SCREEN 2                    INQUIRY MENU

(A)**KEY IN CONTRACT NO. THEN PRESS RESPECTIVE KEY
                CONTRACT NO. ****** ******
        CONTENTS OF CONTRACT DETAILS
        SUPPLIER DETAILS.................................................................PRESS A
        DATES AND TERMS OF CONTRACT ...........................................PRESS B
        TRANSACTIONS...................................................................PRESS 2
        CLIENT NAME, ADDRESS.......................................................PRESS 1
(B)**OR
        INQUIRY BY CLIENT NAME......................................................PRESS 9
        RETURN TO MAIN MENU........................................................PRESS 5
```

Figure 10–11   Inquiry menu 2

There are two methods by which inquiries can be furthered:

- The contract number (Option A)
- The client's name (Option B)

1. If the contract number is known, Option A is used: KEY IN CONTRACT NO.
2. If the client's name is known, Option B is used and the instructions followed: PRESS 9. The result is 'Proceed to INQUIRY BY CLIENT NAME', shown in Figure 10–12. Note that to discontinue an inquiry (whatever panel is on display) it is simply a matter of returning to the inquiry menu or main menu.

```
SCREEN 3
                    INQUIRY BY CLIENT NAME
              KEY IN CLIENT NAME THEN PRESS ENTER
                        *************

INQUIRY BY THIS SCREEN WILL CAUSE THOSE ACCOUNT NUMBERS FOR THE
CORRESPONDING NAME TO BE DISPLAYED

RETURN TO INQUIRY MENU..............................................................................PRESS 9
RETURN TO MAIN MENU..................................................................................PRESS 1
```

Figure 10—12   Inquiry by client name

10.7   REVIEW QUESTIONS

1. (a) Refer to the on-line system case study, Chapter 12. Input media that could be used in this system are:

   - Hand-marked mark sense
   - Hand-lettered OCR
   - Machine-prepared OCR turnaround document
   - VDU

   Outline how each of these media might be used for input of student enrolment data. Comment on the relative suitability of each media.

   (b) If a VDU input media was selected for the physical implementation of the system, what type of dialogue would be most appropriate for the student enrolment procedure? Illustrate your answer by preparing screen layouts for processes 2.1 and 2.2 in Figure 12–3.

2. Prepare a program specification for the edit and update program of Figure 11–20. This specification should be in the form shown in Section 10.6.3, and should contain sufficient detail for writing the program without reference to other documentation.

3. Describe and illustrate the differences between logical and physical design.

4. Using the layout illustrated in Section 11.7.2, and the information in the report layouts of Chapter 12, document the subject file and student file of the on-line system (Chapter 12):

   (a) Indicate the recommended file organization and key(s) in each case

   (b) Will these files contain fixed or variable length records?

5. The logic given in Chapter 12 provides no means of deleting a subject from the subject file. Document modifications to the design to enable this to be carried out.

11 Case Study 1 – A Batch-oriented Systems Design Case

11.1 INTRODUCTION

The objective of this chapter is to illustrate the application of the methodology given in Chapters 8 and 9 to the analysis and design of a real-life application. It has been necessary to choose a rather simple system and to omit some of the details in the documentation in order to limit the size of this chapter.

   The lifecycle concept given in Chapter 1 involves successive passes through the entire system with the cut going to a greater depth of detail with each pass. Thus the complete documentation of a system involving the feasibility study, the logical design and physical design contains much repeated information. It is principally this which has been omitted. As an example the output details which in the course of the project would have been broadly defined in the feasibility study, worked over again in the logical design, and again for layout in the physical design, are presented here only in the logical design. One report is shown in the physical design; it is laid out on a printer spacing chart to illustrate the design and display of printed output.

11.2   THE STAFF CLUB

The Staff Club, an autonomous self-governing body, exists to foster a sense of community and good relations between staff by providing meeting facilities, eating and bar services for staff members. In addition, its facilities are made available for special functions and conference catering. It has a membership of around 650 members who pay $50 membership dues each year. The membership is expected to remain relatively static in the foreseeable future.

A major feature of the Club is its extensive and well-stocked wine cellar, containing over $70 000 value of wine. The wine is sold for consumption in the club or sold on a take-away basis. While a retail price is charged on wines consumed on the premises, members pay a discounted price for take-away. Twice a year wine sales are held at which prices are further discounted to clear excessive or slow moving inventories.

Members may pay cash or use credit facilities for both consumed food and drink, and purchased take-away goods. In the case of food and drink consumed in the club which are to be charged, the member signs the back of the cash register docket. This is then placed in the cash register for later reconciliation of cash and manual posting to the member's account. Most take-away liquor purchases are made on credit and these are handled with a slightly different system using a wine-order charge book kept in the bar. The member's name and order details are written in this book at the time of ordering. Each month the pages of the book for the collected orders are torn out to prepare the members' statements. No credit limit is applied to members' accounts and there have been very few bad debts. However, if the account is not paid in the 30-day account period the discount given on the take-away liquor purchases is forfeited.

The club is organized with overall responsibility vested in an elected committee headed by a president. A manageress, reporting to the committee, bears executive responsibility and supervises the ten regular employees of the club. They are:

> 2 cooks
> 5 waitresses (3 part-time)
> 1 bar attendant
> 1 office clerk
> 1 storeman (part-time)

All the club's office systems are at present manual and, given the modest scale of most of them, operate quite successfully. The major problem area is controlling the large wine inventory which has in excess of 50 000 bottles. Since the holding of beer and spirits is quite small, these areas present no difficulty.

11.3   REVIEW OF CURRENT SYSTEMS

This review concentrates on the wine purchasing, receiving and sales system, as wine constitutes the principal investment of the club and its control constitutes the major problem area. The cash reconciliation and accounting systems are briefly covered. The short descriptions below should be read in conjunction with the flowcharts in Figures 11–1 to 11–9.

Wine purchasing (Figure 11–1). The decision to purchase wine may arise in one of three ways:

(a) The club Vice-President, as the wine connoisseur, is responsible for ordering new wine lines for the cellar within an approved budget.

(b) If a member wishes to purchase a wine in quantities of 1 dozen or more, a special order is made.

(c) In preparation for a special function an order may be prepared.

All wine purchase orders are recorded in the Wine Order Book.

Wine receipts (Figures 11–2 and 11–3). Following receipt of a shipment of wine the storeman checks the physical receipts against the supplier's invoice and marks any discrepancies. The manageress then checks the invoice against the order in the Wine Order Book and notes the receipt. If the order is for the cellar (i.e. part of the Vice-President's order), a costing card for that wine type is updated with the new actual cost paid. If the shipment represents a new wine type no costing card will exist and one is prepared. The costing card contains:

> Wine description and vintage
> Date received
> Quantity received
> Wholesale cost/dozen
> Retail price (non-member price)
> Discount price
> Bin number (location in cellar where wine is stored)

The retail price is the current recommended retail price. The discount price is calculated by including a 7 per cent profit margin on the wholesale cost after allowing for government taxes at a fixed percentage rate.

Wine purchased for a special function or to fill a special member's order does not enter the cellar inventory. The flowchart gives further details for these two categories of purchase.

Wine sales with meals (Figure 11–4). The wine is either paid for in cash or charged to the member's account by signing the cash register docket.

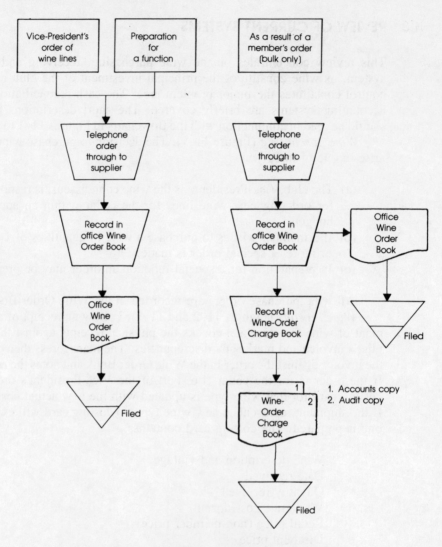

Figure 11—1   Wine purchasing flowchart

Wine price revaluation (Figure 11–5). Many of the wines purchased by the club are intended to mature in the cellar for some years before they will be recommended for purchase and consumption. The value of the wine increases over time and to reflect this the club annually revalues its holding. Two sources are used for the revaluation: Thompson's Liquor Guide and Gray's Auctions. Revaluations are also performed (see Figure 11–6) at the manageress's discretion during the course of the year.

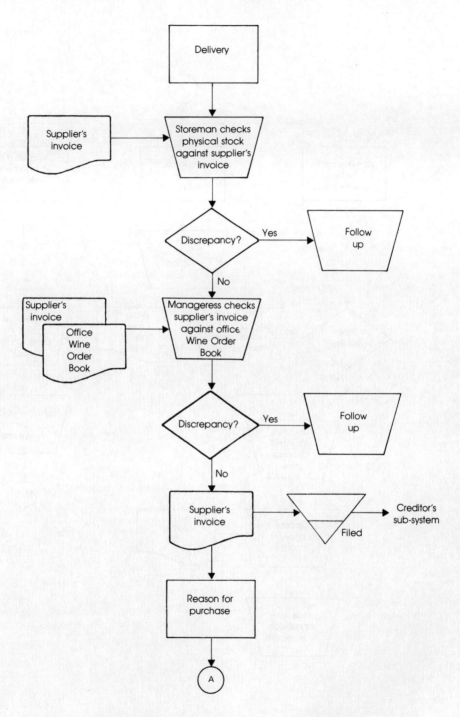

Figure 11—2 Stock receipts flowchart

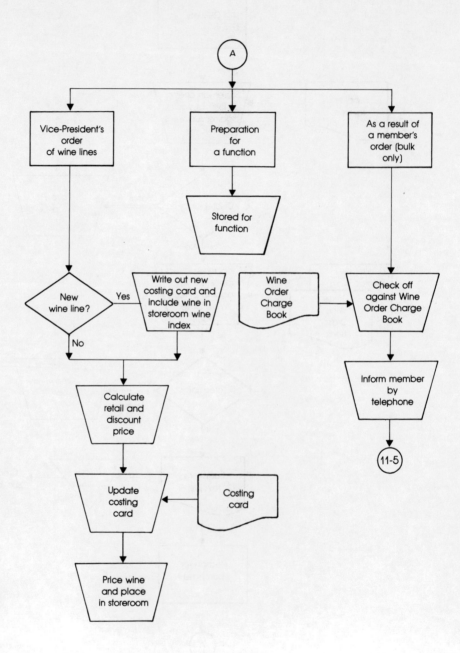

Figure 11—3   Stock receipts flowchart (continued)

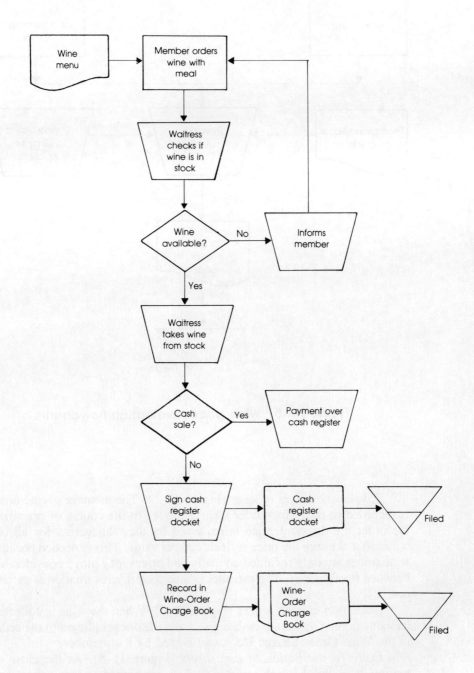

Figure 11—4   Wine sales with meals flowchart

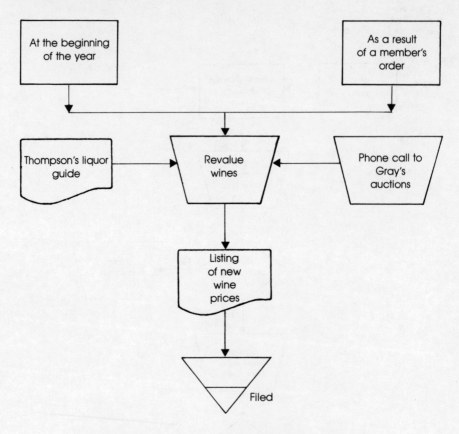

Figure 11—5   Wine price revaluation flowchart

Take-away orders of wine (Figure 11–6). The member's wine order is recorded in the Wine-Order Charge Book. In the course of preparing the order the discount price is reviewed by the manageress for all old wines to make sure the price reflects current value. This is needed because wine prices are only revalued annually and prices may move considerably between revaluations. The member is informed if a revaluation is carried out.

Collection of wine orders (Figure 11–7). When the wine is collected its value is rung up on the cash register and the docket pinned to the order in the Wine-Order Charge Book and signed by the member.

Daily reconciliation of day sheets (Figure 11–8). At the close of trading each day the cash register dockets are pinned to a day sheet and reconciled to the total of cash and credits.

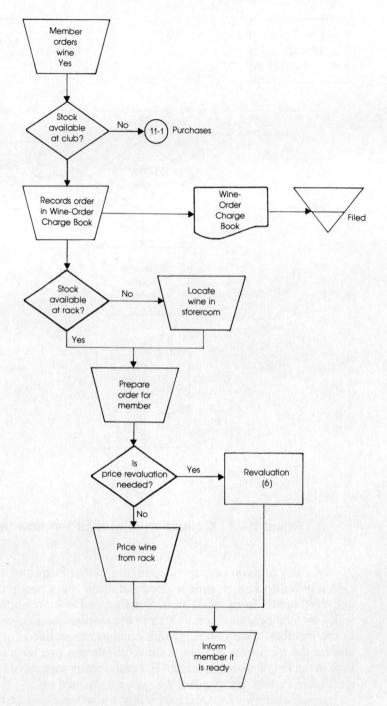

Figure 11—6   Take-away wine orders flowchart

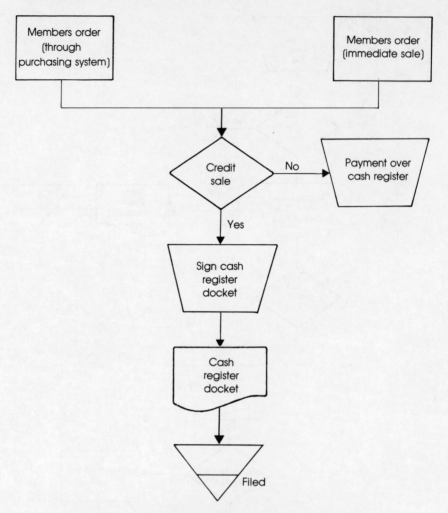

Figure 11—7   Collection of wine orders flowchart

Monthly account preparation (Figure 11–9). Each month the signed cash register dockets relating to food and drink consumed in the club and delivered orders from the Wine-Order Charge Book are sorted alphabetically by member, and then written to the monthly account which is sent to the member. Two copies of this account are retained for office use, one for the member's accounts receivable file and one for a monthly file kept in docket number sequence. The cash receipt aspects of the accounts receivable system are standard and not commented on.

Physical stocktake. Quarterly a physical stocktake is carried out and the contents of the cellar are listed and the current value computed.

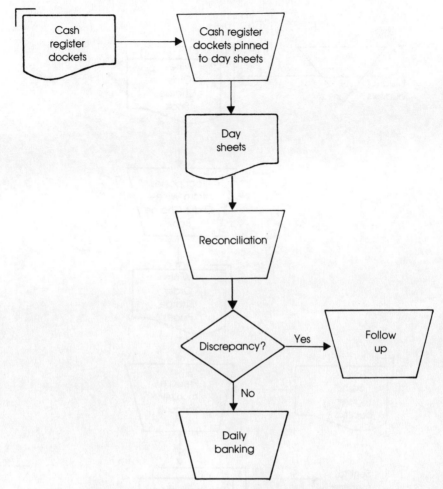

Figure 11—8   Daily reconciliation flowchart

11.4   SHORTCOMINGS OF CURRENT SYSTEM AND MANAGEMENT REQUIREMENTS

The major shortcoming of the current system perceived by the club's management is the lack of control over the wine inventory. Since no perpetual stock records are kept there is no way of knowing the holding of a particular wine without going to the cellar, physically locating it using the Storeroom Wine Index and counting the number of bottles on hand. It is only following stocktakes that the club has a record of its stock.

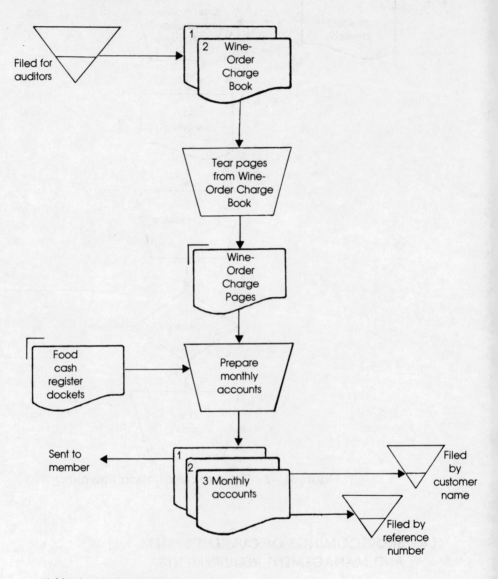

1. Member's copy (white)
2. Accounts receivable copy (green)
3. File copy (pink)

Figure 11—9   Monthly account preparation flowchart

Another aspect of this lack of inventory control is that the club is only able to calculate its profit after the completion of a stocktake. A loss has been made in the past two years which has been attributed to (a) a failure to review food prices owing to a lack of timely knowledge of the trading profit and loss situation, and (b) a failure to adequately market the wine due to no up-to-date lists of stocks and prices.

A computerized inventory and accounting system is viewed by the club management as a possible solution to their difficulties. But wine inventory control is considered to be the priority area since an improved accounting system, with monthly reporting, cannot function accurately without either keeping a perpetual inventory value or carrying out monthly stocktakes which would be prohibitively expensive. Thus management requested a feasibility study to be carried out on a wine inventory system with provision made for a simple interface to an accounting and accounts receivable system that will be considered later. The defined objectives of the system are for the production of management information to aid in the control of the wine inventory and to include reports covering:

- Sales and profitability of wines by wine type
- Stocktake listing
- Stock of wine by wine type and wine origin
- Wine lists for distribution to members

Details of the required data fields in these reports are to be organized by wine type and within wine type by wine origin. Wine types are red, white and sparkling, and wine origins are local and imported. Within each wine type are standard bottles and bulk, where bulk includes all quantities in excess of standard bottles such as wine casks and 2-litre bottles.

11.5 ALTERNATIVES EXAMINED IN FEASIBILITY STUDY

Perhaps the most critical aspect of a computer-based system for the wine inventory is determining the most appropriate means for the correct identification and input of wine sales. This is particularly a problem for wine consumed in the club (25–30 bottles per day) and small take-away purchases. The staff are generally so busy during the lunch hour that insufficient time is available to record correctly the wine description and vintage unless a simple trouble-free and rapid means is devised. The alternatives selected for consideration were:

1. Collect the empty bottles for wine consumed in the Club and record details after the lunch rush. Record manually the details of all other purchases.

   This idea was abandoned as members may decide to keep the bottle if it is not completely empty or perhaps just as a memento.

2. Manually record all details on a bar chit for later entry to the system.

   This alternative was not accepted as it was believed the accuracy of input data would not be adequate and it would take too much time.

3. Employ an adhesive-backed label that can be peeled off a bottle, bulk container or case and fastened to a card. The procedures to be adopted for the various categories of sales being:

   (a) Dining area — labels are removed from bottles and fastened to a card to record all sales that day in this area. Sales of bulk wine by glass or carafe are accounted for by peeling off the label of the bulk container when it is first opened.

   (b) Small orders (single bottles) take-away or bar — a similar procedure to that followed in the dining area will be used.

   (c) Orders for case lots — the order can be entered in a book similar to that used currently and when the cases are assembled the outside case label is removed and fastened to the order. This also indicates the quantity supplied, which may on occasion differ from the quantity ordered.

   This alternative sees both a bottle label and a case label being produced (possibly by the computer) on receipt of a case. The case label is affixed to the case with the bottle labels attached to the front of the case in such a way that if the case is opened later they can be fastened to each bottle.

In addition to alternatives for data capture three alternatives for the system were considered:

1. Maintain the manual system with some redesign

2. Implement a computer system on the main computer of the organization using a VDU for interactive input and a small local-printer for short reports. Long reports would be produced off-line on the main-line printer

3. Implement a computer system on a microprocessor, with its own VDU, floppy disks and printer

Brief designs for each of these alternatives were sketched out and costs and benefits estimated. On the basis of this analysis the Club's management decided to adopt the label approach to data capture and to use the mainframe computer of the organization.

In considering the aspect of the detail of information required, management considered two alternatives:

1. Confine the system to keeping a perpetual inventory and reporting stock value and costs of goods sold.
2. Have the system also report on the profitability of each wine, and by addition the profit of each wine type and the total profit.

To provide the capability described in the second alternative would require the value of each sale to be captured and entered to the computer. It would not be sufficient to assume that either the member price or non-member price applied. This is because of such exceptions as promotional offers, special sales and wine provided free for club committee dinners. These exceptions could be covered by extending the number of prices carried in the file and entering the code corresponding to the appropriate price or alternatively entering the sales value for each transaction.

The systems analyst advised the club to adopt the first alternative for the following reasons:

- Being a simpler system alternative (1) would place less demand on staff
- The club is a first-time user so a simple solution should be adopted
- Total club profit could be calculated using alternative (1)

Management adopted this recommendation, although they were disappointed at the prospect of not getting the profit for each wine and wine type. The systems analyst pointed out that a future enhancement of the system to provide profit by wine could be carried out if experience with the system suggests it is worthwhile.

11.6 LOGICAL DESIGN

11.6.1 Content of Management Outputs

The system will produce the following management reports and information. Outputs related to control of the system are covered in a later section.

Wine lists

These are prepared for distribution to members, probably on a monthly basis. The information to be included is:

> Wine type
> Wine origin
> Description
> Vintage
> Stock number
> Member price
> Non-member price
> Bin number

The report will be sorted by wine type, wine origin, description and vintage.

Labels

Labels are produced for both cases and bottles with the bottle labels only affixed when the case is broken. The labels are printed as soon as possible after the receipt of the wine and carry the wine stock number which provides a simple and positive identification for each wine for data capture purposes.

The data printed on each label is:

> Stock number
> Description
> Vintage
> Bin number
> Quantity (if a case label)

The bin number is the location in the cellar where the wine will be kept. It is given to the wine by the cellarman when the wine is initially received and checked. The stock number is a unique identifying numerical code for each district wine/vintage/bottle size.

Stocktake sheets

The purpose of this report is to list the wine description and vintage ready for the recording (on this list) of the counted physical stock. It is sorted in bin number/stock number sequence to help the stocktaker.

The report contains:

Bin number
Stock number
Description
Vintage
Blank column headed physical quantity for completion by stock-taker

Stock and sales report

This monthly report to management gives the stock and sales units of each wine. Special orders purchased 'on indent' for a customer are all reported as a single category under wine type 'special orders'. Orders purchased for special functions are reported under the wine type 'special functions'. See wine type in data dictionary for codes used. The information included is:

Wine type
Wine origin
Description
Vintage
Stock number
Bin number
Total quantity purchased
Quantity on hand
Revalued cost per unit
Historical average cost per unit
Member price
Non-member price
Quantity sold this month
Cost of sales this month (at revalued average cost)
Quantity sold year-to-date
Costs of sales year-to-date
Stock loss units
Stock loss cost

Costs are kept for historical average cost and revalued cost. Most of the quality wines are purchased once at release of the vintage and each year their cost is revalued to reflect the extra value they have accrued through greater age. This greater value of inventory is then taken up in the accounting system. At times for quality wines and frequently for bulk type wines, additional purchases of the same wine are made. In this event the cost of the new stock is averaged with that currently held. This is done for both the revalued and the historical cost.

The member price is calculated at a given percentage increase on the revalued cost while the non-member price is the current retail price.

Even though most wines are sold by the dozen case the unit of measure for all wine is the single bottle, cask or container. This has been done to avoid confusion.

Stock loss records the loss since the start of the year arising from breakage, spoilage, or stocktake discrepancy, and is costed at the revalued cost.

11.6.2 Input Documents

Stock purchase advice

The purchase advice input data is drawn largely from the supplier invoice that has been marked off by the storeman to show actual quantity received and bin number allocated and checked off by the manageress against the wine order book. This allows the wine type to be determined. (The wine type is used to indicate a special order or a special function order and this information is determined by reference to the wine order book.) The actual data input to the system can be seen by referring to 'purchase advice' in the data dictionary. This assumes the wine is a new purchase.

If the wine is already in stock only the following data need be input:

> Stock number
> Quantity received
> Total cost
> Non-member price

Sales advice

The sales advice is the system input relating to any sale or other issue of stock. Other issues may include promotionals, free wine for special visitors and club gifts. The data items on the sales advice are:

> Stock number (from label pinned to advice)
> Quantity (on label if case lot)

Stock adjustment

Stock adjustments may relate to stocktake discrepancies (stock shrinkage), breakage and spoilage. A separate code is used to designate each category. Input data is:

> Stock number
> Adjustment code
> Quantity (+ or −)

Stock master file maintenance

Any field on the stock master file may need to be changed but the most commonly occurring change will be due to revaluations of stock cost and non-member (retail) price. The input data may comprise the set of data items on the stock master file that need to be modified.

Order book

The order book is used to record all details of a customer's (member or non-member) order of wine. It remains the same in the new computer system as it did in the manual system. Consideration was given in the feasibility study to having this file handled by the computer but this was rejected as it was felt it could only be successfully computerized with an on-line system and this alternative was not feasible. The order book contains:

> Order number
> Member details — member name, contact phone number
> Date
> Wine order details:
> > Regular order/special order/special function
> > Wine description
> > Vintage
> > Quantity ordered
> > Price quoted
> Total value of order

The order book does not form a direct input to the computer system but forms a key link in the total stock control system as does the supplier order book.

Supplier order book

The supplier order book is used to record all wine orders placed on suppliers. Its data includes:

> Order number
> Supplier details (name, address, etc.)
> Wine order details:

It will remain a manually kept book in the new system and interacts only indirectly with the new computer system.

11.6.3 Data Flow Diagram — Logical System

The data flow diagrams in Figures 11–10 to 11–13 include some manual processes in order to indicate the source of input data. Where appropriate the man/machine boundary as defined in the feasibility study is shown.

11.6.4 Data Dictionary

This is organized in two parts. The first part (a) contains descriptions of the data flows and files in the system, while part (b) contains the code structures.

(a) Data flow and file descriptions

Customer Order:
    Order number
    Customer name, address and phone number
    Date
    Member/non-member indicator
    Order line:*
        Stock number
        Description
        Vintage
        Quantity

Delivery Check:
　　Description
　　Vintage
　　Quantity
　　Bin number
File Details:
　　Stock number
　　Unit price
　　Quantity on hand
　　Bin number
Marked off Invoice:
　　Supplier invoice:
　　(Quantity received
　　Bin number)*

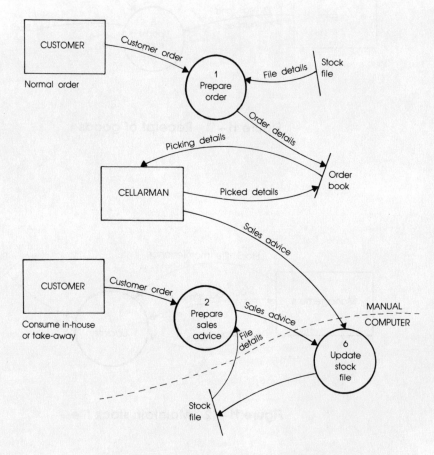

Figure 11—10   Customer order

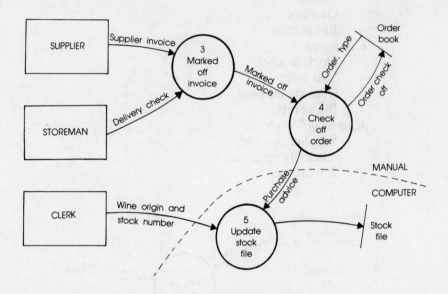

Figure 11—11   Receipt of goods

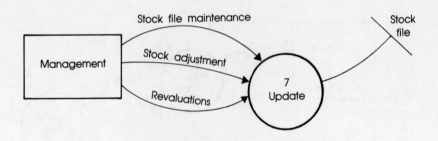

Figure 11—12   Maintain stock file

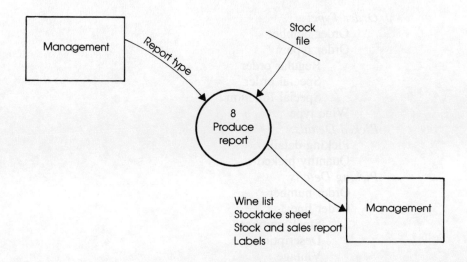

Figure 11—13   Produce reports

Order Book:
    Order details:
Order Check Off:
    Order number
    Order line received:*
        Description
        Vintage
        Quantity received
Order Details:
    Order number
    Customer name, address and phone number
    Date
    Member/non-member indicator
    Completed order line:*
        Stock number
        Description vintage
        Bin number
        Unit price
        Quantity
        Extended price
    Total value

Order Type:
Order number
Order type:*
Regular order
Special order
Special function
Wine type
Picked Details:
Picking details
Quantity picked
Picking Details:
Order number
Order line:*
Stock number
Description
Vintage
Bin number
Quantity
Purchase Advice:
Description
Vintage
Wine type
Bin number
Quantity
Unit cost
Recommended retail price
Total cost
Revaluation:
Stock number
Revalued unit cost
New recommended retail price
New recommended member price
Sales Advice:
Date
Sales transaction:*
Stock number
Quantity

Note: The label which forms the sales advice input also contains
the description, vintage and bin number. However, as only the
stock number need be input to identify the wine, the description
vintage and bin number are omitted from the data dictionary.

Stock Adjustment:
Stock number
Adjustment
Reason code
Stock File:
Stock number
Wine description
Vintage
Bin number
Wine type
Wine origin
Quantity on hand
Total quantity purchased
Average unit cost
Member price
Non-member price
Month-to-date sales units
Year-to-date sales units
Revalued unit cost
Month-to-date cost of sales
Year-to-date cost of sales
Stock loss units
Stock loss cost
Stock File Maintenance:
Stock file
Supplier Invoice:
Supplier name and address
Invoice number
Date
Club order number
Invoice line:*
Description
Vintage
Quantity
Unit price
Recommended retail price
Extended price
Total value
Discounts
Net value

(b) Coding structures

Bin code: three-digit non-structured code
Stock adjustment code: one-digit code to designate reason for stock adjustment
    1 = stocktake discrepancy
    2 = breakage
    3 = spoilage
Stock number: six-digit non-structured code
Wine origin: one-digit code to designate local or imported wine
    1 = local wine
    2 = imported wine
Wine type: one-digit code to designate wine type, bottle size and special order or function
    1 = red standard bottle
    2 = red bulk
    3 = white standard bottle
    4 = white bulk
    5 = sparkling standard bottle
    6 = sparkling non-standard bottle
    7 = special order
    8 = special function
    9 = fortified
    0 = spirits

11.6.5   System Control

This section specifies the broad strategy to be used in ensuring the integrity of data processed by the system. Details of how the strategy is implemented are left to physical design.

The sales input will be controlled by batching sales advices and balancing the relevant input data to the figure given on the batch slip. Since it is quite unlikely that there will be more than one purchase advice or stock adjustment form on any one day, these documents will not be batched. At stocktake time there will doubtless be many stock adjustment forms but it does not seem warranted to bear the extra effort of batching single documents for this quarterly event. However, to improve data integrity a hash total of adjustment quantity is included on the bottom of the stock adjustment form. Manual control methods will also be used for this occasion. Note that this control information has been largely omitted from the data flow diagram and the data dictionary so as not to obscure the major functions of the system.

(a) *Sales advice batch header*
   This will contain:
      Date
      Batch number
      Number of sales advices
      Hash total of quantity
   The batch number will be allocated sequentially from zero each day.

(b) *Sales edit control report*
      Date, batch number
      Total sales advices read and quantity hash total read
      Number of sales advices and hash total from batch header
   As input will be keyed and edited interactively, edit rejections are cleared up before proceeding and this report is for control only.

(c) *Edit reports for purchases, stock adjustments, revaluations and maintenance*
   These reports will list all the input data and present a hash total on quantity.

(d) *Sales update report*
      Date
      Brought forward hash total quantity of stock
      Brought forward total value of stock
      Transactions hash total quantity
      Transactions costed total value
      Carried forward hash total of stock
      Carried forward total value of stock

(e) *Purchases update report*
      Date
      Brought forward hash total quantity of stock
      Brought forward total value of stock
      Transactions hash total quantity
      Transactions costed total value
      Carried forward hash total of stock quantity
      Carried forward total value of stock

(f) *Stock adjustment, revaluation and sundry maintenance update report(s)*
   As for sales update report but the report(s) are to be segmented by wine type and wine origin.

11.7 PHYSICAL DESIGN

While the major elements of the physical design are presented in this section not all aspects have been documented. For example only one

VDU screen layout and one printer output layout are given as examples. In addition, a number of design aspects such as the security and implementation plan have been omitted from this case in order to focus attention more clearly on the basic elements of analysis and design.

11.7.1 General Sequence of Runs

Although the sequence of the update operations were not discussed in the logical design (the DFDs show no aspects of timing), it is clear that there should be one. Broadly the stock file maintenance needs to be carried out first, then the receipts from suppliers added to the file before club sales of wine are subtracted.

11.7.2 Stock File Layout and Organization

Key: stock number

Name	Description	Specification
STNO	Stock number	9(6)
WINES	Wine description	X(30)
VIN	Vintage	9(2)
BIN	Bin code	9(3)
W-TYP	Wine type	X
W-ORG	Wine origin	9
Q-O-H	Quantity on hand	9(4)
T-Q-P	Total quantity purchased	9(5)
AVCST	Average unit cost	99.99
MEMP	Member price	99.99
N-MEMP	Non-member price	99.99
MTDSALU	Month-to-date sales units	9(4)
YTDSALU	Year-to-date sales units	9(5)
REVCST	Revalued unit cost	99.99
MTDCST	Month-to-date cost of sales	9(5).99
YTDCST	Year-to-date cost of sales	9(6).99
STKLOSSU	Stock loss units	9(3)
STKLOSSV	Stock loss cost	9(4).99

Stock file organization

The obvious access strategy for the stock file is direct access, as the file processing is characterized by a large file with a low activity rate, and a moderately low volatility.

In considering the alternatives of random, index sequential and direct addressing, the following factors appeared relevant:

- Direct addressing would provide the fastest access, followed by random and then index sequential.

- Since the keys are allocated sequentially, initially all posible keys will correspond to products. However, as the products are sold out key gaps will occur. These keys could be re-assigned later to new products when the history held on the stock file for the obsolete product was no longer needed. Note that control will need to be exercised over the deletion of stock file records.

- A listing in key sequence is not called for in the specification.

On weighing up this decision on file organization it was decided to adopt a direct addressed file with provision to be made for allocating obsolete keys to new products. The key would initially be allocated sequentially starting from 0001.

11.7.3 Input Documents

The logical design shows the following inputs are required to the computer system:

> Stock adjustments
> Sundry maintenance
> Revaluations
> Purchases advice
> Sales advice

While input for the stock adjustment, file maintenance and revaluations could be performed using one input form, it was decided that control would be improved if three separate forms were used. The general purpose 'sundry maintenance form' is only used when the maintenance is not a revaluation or stock adjustment. The sundry maintenance form is used to delete a record but as this action erases its history, it should only be taken after all reporting has been carried out.

Sketches for three of the input forms are given in Figures 11–14 to 11–16.

```
┌─────────────────────────────────────────────────────────────────┐
│                                                                   │
│                          STAFF CLUB                               │
│                    STOCK ADJUSTMENT ADVICE                        │
│                                                                   │
│   Input Form No. 5                        Sequence No. 9999       │
│                                           Date......./......./........ │
│                                                                   │
│  ┌──────────┬────────────────┬────────────┬──────┬──────┐        │
│  │          │                │ Adjustment │ Sign │Reason│        │
│  │Stock number│  Description  │  quantity  │+ or −│ code │        │
│  ├──────────┼────────────────┼────────────┼──────┼──────┤        │
│  │          │                │            │      │      │        │
│  │          │                │            │      │      │        │
│  │          │                │            │      │      │        │
│  │          │                │            │      │      │        │
│  │          │                │            │      │      │        │
│  │          │                │            │      │      │        │
│  │          │                │            │      │      │        │
│  │          │                │            │      │      │        │
│  ├──────────┼────────────────┼────────────┼──────┼──────┤        │
│  │          │  HASH TOTAL    │            │      │      │        │
│  └──────────┴────────────────┴────────────┴──────┴──────┘        │
│                                                                   │
│  Unit of measure for quantity is bottle, cask or flagon.          │
│                                                                   │
│  Reason code                                                      │
│  1. Stocktake discrepancy                                         │
│  2. Breakage                                                      │
│  3. Spoilage                      ...............................  │
│                                           Approved                │
│                                                                   │
└─────────────────────────────────────────────────────────────────┘
```

Figure 11—14   Stock adjustment advice input form

STAFF CLUB
SUNDRY MAINTENANCE FORM

Input Form No. 6

Sequence No. 9999

Date......../......../........

☐ Change a record: Fill in only stock number and fields to be changed

☐ Delete a record: Note: This also deletes its current history

Stock number   Wine description   Vintage   Bin no.

☐ Wine type   ☐ Wine origin $ Average unit cost $ Revalued unit cost

$ Member price $ Non-member price   Total quantity purchased

Month-to-date sales units $ Month-to-date cost of sales   Stock loss units $ Stock loss cost

Year-to-date sales units $ Year-to-date cost of sales   Quantity returned $ Credit amount

Figure 11—15 Sundry maintenance input form

STAFF CLUB
STOCK REVALUATION ADVICE

Input Form No. 7   Sequence No. 9999

                                                        Date......../......../........

Stock No.	Description	Unit cost	Non-member price	Member price

Note: The 'member price' will be calculated using
the standard mark-up if field left blank.

                                        ....
                                        Approved

Figure 11—16   Stock revaluation input form

11.7.4   VDU Screen Layouts

As an example of the VDU screen layouts that will be used with this
system Figure 11–17 illustrates the VDU layout for the stock adjustment
data input.

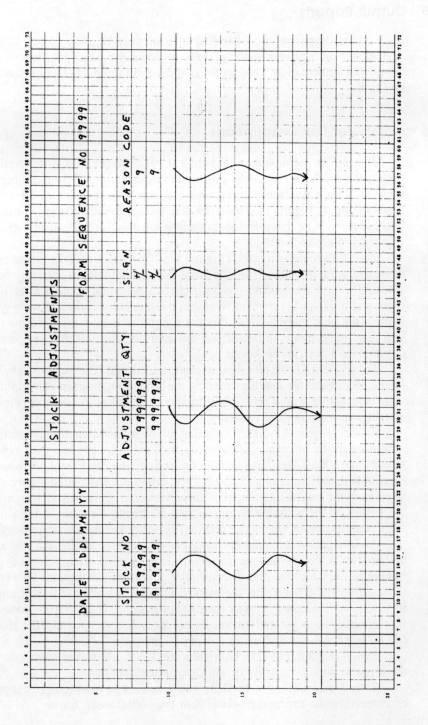

Figure 11—17   Stock adjustment screen layout (VDU)

11.7.5 Output Reports

Each of the output reports specified in content in Section 11.6.1 has to be laid out on a printer spacing sheet to detail the exact location of all the printed information. One report is shown in Figure 11–18 to illustrate this task.

11.7.6 Computer Run Descriptions

Run 1 File maintenance (Figure 11–19). The first run is required to update the stock file for stock adjustments, revaluations and sundry maintenance. This run is really three separate runs, one for each of the three types of input. Figure 11–19 shows the system flowchart used for each of these runs. Note that different programs are used for the edit and update with each class of input.

Run 2 Purchases update (Figure 11–20). Run 2 accepts either purchases of wine not already on the stock file or additions to currently held stock. The edit and update program checks the stock number after it is input and displays description, vintage, bin number, origin and wine type if the wine is already held. If the stock number is new this information will need to be input. In both cases the quantity, the unit retail price and the total cost need to be keyed.

Subsequent to completing the input, the special label paper is loaded on the printer and the labels printed by the label print program. An equally valid alternative would have been to spool a label print file and print it after the receipts control report.

Run 3 (Figure 11–21). The sales advices are batched and totalled according to the principles outlined in the logical design; and then keyed through the VDU. The data is edited on-line and the batch total previously hand-calculated is compared against the computer-calculated figure derived from the transactions input. If this checks out, the edited transactions are written to the edited sales transaction file ready for input to the stock file update program where the relevant quantity fields will be updated.

An alternative arrangement would have been to combine the edit and update functions in the one program. However, if each sales advice transaction updated the file directly after it was edited then in the event the batch total did not agree, the incorrect update would have to be found, reversed and corrected. It was considered easier and less error prone to first prepare the input and ensure its accuracy before updating the stock file on a batch basis and using a separate program.

Run 4 (Figure 11–22). This sorts the stock file to the sequence of the reports: wine type/wine origin/description and vintage.

Runs 5 and 6 (Figure 11–22). These runs print the required reports by extracting the appropriate data from the sorted stock file.

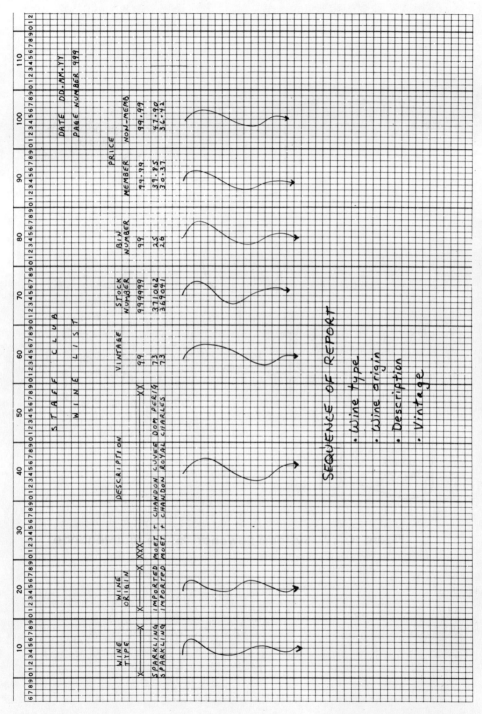

Figure 11—18   Printer layout for wine list

RUN 1

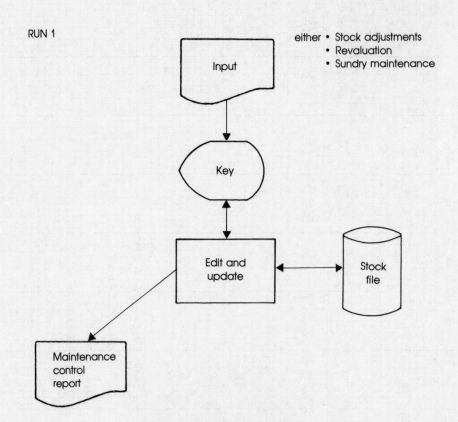

Figure 11—19 Systems flowchart for file maintenance

RUN 2   EDIT AND UPDATE PURCHASE RECEIPTS

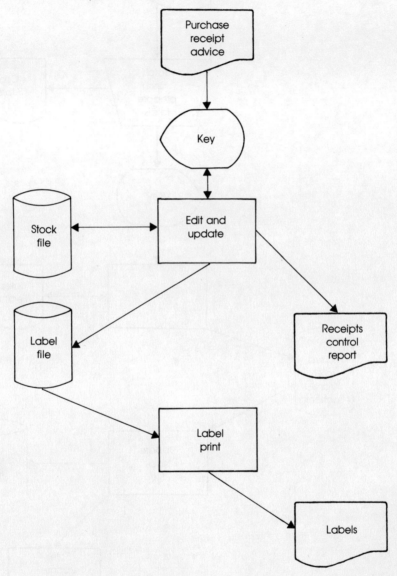

Figure 11—20   Systems flowchart: Run 2

RUN 3

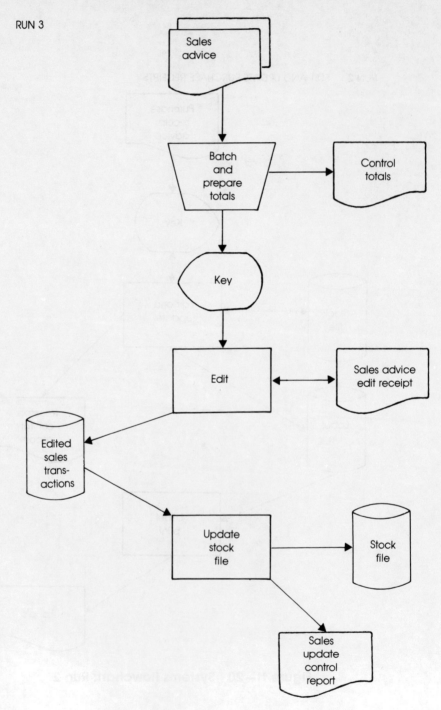

Figure 11—21   Systems flowchart: Run 3

RUN 4   PREPARE FOR PRINTING REPORTS

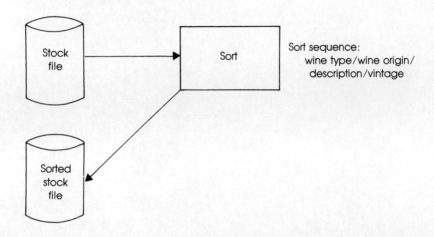

RUN 5   PRINT MONTHLY REPORT

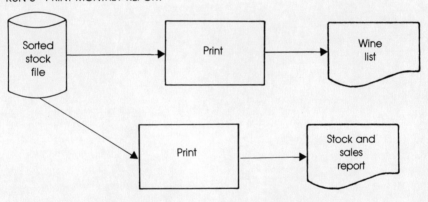

RUN 6   PRINT STOCKTAKE REPORTS

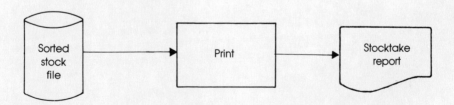

Figure 11—22   Systems flowchart: Runs 4 to 6

12 Case Study 2 – An On-line System

The following documentation describes a university student enrolment system. It is a terminal-based real-time system serving two user groups: students and faculty administration. This is shown in Figure 12–1, the system context diagram. The inputs to the system are enrolment details from students, and subject details and report requests from the faculty. The outputs are reports, enrolment forms, and responses which are made during terminal operation. The system is described by way of a set of data flow diagrams, process definitions, and a data dictionary. The student will observe that the process definitions are written in a structured form of English, similar to the pseudocode in Chapter 3.

The documentation provided in this chapter outlines a logical design for a system which will allow graduate students in the masters program to complete their enrolment procedures interactively by providing data to the system at the beginning of each academic session. The faculty administrative staff will use the system to gain information on class sizes, persons enrolled in classes, and persons claiming special approval to enrol in subjects which would otherwise be unavailable to them.

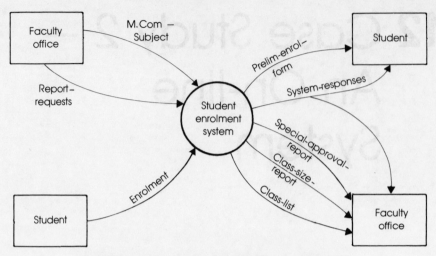

Figure 12—1   System context diagram

Figure 12–2 shows a partitioning of the system into its major functional parts. These are:

- PROCESS 1 Set up M.Com subjects
- PROCESS 2 Enrol student
- PROCESS 3 Extract reports (Processes 2 and 3 are further partitioned in Figures 12–3 and 12–5)

Note that many of the system data flows (e.g. as shown in Figure 12–3) are not present in this figure. Each level of data flow diagram shows the analyst's perception of the major parts and flows in the system when viewed at that level of detail. At the lowest level the reader can obtain the complete detail of the flows in the system. For example only one flow (Subject-updates) goes from process 2 to the subject file, but to complete this update many flows are necessary; see Figures 12–3 and 12–4.

Process 1 occurs at the beginning of each year to set up the subject-file for that year. Thus faculty staff input details are listed in the data dictionary for each subject so that the file can be created. The logic for process 1 is as follows:

For each M.Com SUBJECT
If UPDATE-TYPE = 'D' (delete) then
— accept the SESSION, SUBJECT-NO and CLASS-TUT-
CODE from the terminal

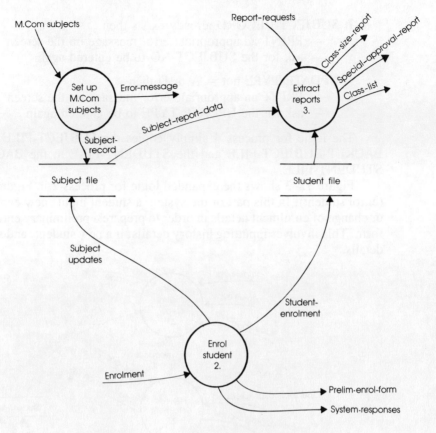

Figure 12—2   Student enrolment system

      — find the SUBJECT-RECORD in the SUBJECT-FILE
If found then
      — If NO-STUDENTS-ENROLLED-IN-CLASS and NO-
        STUDENTS-ENROLLED-IN-TUT are zero then
        — DELETE the SUBJECT-RECORD otherwise (subject
        not found)
      — report an error

If UPDATE-TYPE = 'A' (add) then
      — accept all MCOM-SUBJECTS fields from the terminal
      — validate input fields as per specifications
If any validation errors occur then
      — PRINT an appropriate error message on the screen
      — ask for the field to be entered again, otherwise
      — insert a SUBJECT-RECORD into the SUBJECT-FILE

If SUBJECT-RECORD already exists then
— PRINT an appropriate error message on the screen
— ask for the SUBJECT-NO to be entered again

If UPDATE-TYPE not = A or D then
— PRINT an appropriate error message on the screen
— ask for the UPDATE-TYPE to be entered again

The logic for process 4 simply copies the SUBJECT-FILE to the BACKUP-SUBJECT-FILE and the STUDENT-FILE to the BACKUP-STUDENT-FILE.

Figure 12–3 shows the expanded logic for process 2 in Figure 12–2 (enrol student). In this part of the system a student inputs new enrolment or change of enrolment details in order to prepare a preliminary enrolment form. This involves inputting history details, if a new student, and subject details.

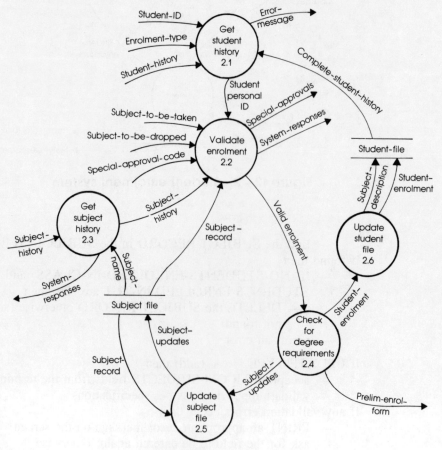

Figure 12–3   Enrol student

Get student history (process 2.1)

Get the STUDENT-ID from the terminal
Carry out a check digit verification on the STUDENT-NO
    If the STUDENT-NO is in error then
        — display an appropriate error message on the terminal
        — ask for the STUDENT-NO to be entered again
    If ENROLMENT-TYPE = 'N' (new enrolment) then
        — check that the STUDENT-NO does not already exist
          on the STUDENT-FILE
        — get the STUDENT-HISTORY from the terminal
        — validate each STUDENT-HISTORY field as per speci-
          fications
    If any fields are found to be in error
        — display an appropriate error message on the terminal
        — ask for the field to be entered again
    If ENROLMENT-TYPE = 'C' (change of enrolment) then
        — get the COMPLETE-STUDENT-HISTORY from the
          STUDENT-FILE using STUDENT-NO
    If the STUDENT-RECORD does not exist on the STUDENT-
FILE
        — display an appropriate error message on the terminal
        — ask for the STUDENT-NO to be entered again
    If ENROLMENT-TYPE not = 'N' or 'C' then
        — display an appropriate error message on the terminal
        — ask for the ENROLMENT-TYPE to be entered again

Get subject history (process 2.3)

Get EXEMPT-SUBJECT from the terminal
Find the SUBJECT-RECORD in the SUBJECT-FILE using
SUBJECT-NO
If the SUBJECT-RECORD does not exist then
    — display an appropriate error message on the terminal
    — ask for the EXEMPT-SUBJECT to be entered again
Display the SUBJECT-NAME on the terminal
Ask if the EXEMPT-SUBJECT is OK
If not
    — go back to the start of the process
Allow a maximum of four EXEMPT-SUBJECTS to be entered
Get SUBJECT-ALREADY-PASSED from the terminal
Find the SUBJECT-RECORD in the SUBJECT-FILE using
SUBJECT-NO

If the SUBJECT-RECORD does not exist then
— display an appropriate error message on the terminal
— ask for the SUBJECT-ALREADY-PASSED to be entered again

Display the SUBJECT-NAME on the terminal
Ask if the SUBJECT-ALREADY-PASSED is OK
If not
— go back to the start of the process
Allow a maximum of ten SUBJECT-ALREADY-PASSED to be entered

Check for degree requirements (process 2.4)

For each VALID-ENROLMENT check
— (a) that at least 4 CORE-UNIT SUBJECTS ARE PRESENT
— (b) that not more than 6 CORE-UNIT SUBJECTS ARE PRESENT
If any errors then
— reject the entire enrolment
— PREPARE A PRELIM-ENROL-FORM as per sample report layout

Update subject file (process 2.5)

For each SUBJECT-TO-BE-TAKEN
— find the SUBJECT-RECORD in the SUBJECT-FILE
— UPDATE the NO - STUDENTS - ENROLLED - IN - CLASS and
  NO - STUDENTS - ENROLLED - IN - TUT
— UPDATE the ENROLLED-STUDENT table and the TUT-ENROLLED-STUDENT table
For each SUBJECT-TO-BE-DROPPED
— find the SUBJECT-RECORD in the SUBJECT-FILE
— downdate the NO-STUDENTS-ENROLLED-IN-CLASS and
  NO - STUDENTS - ENROLLED - IN - TUT
— REMOVE the STUDENT-NO from the ENROLLED-STUDENT table and from the TUT-ENROLLED-STUDENT table

Update student file (process 2.6)

> For each STUDENT-ENROLMENT
> > If ENROLMENT-TYPE = N (new enrolment) then
> > — build a STUDENT-RECORD
> > — insert the STUDENT-RECORD into STUDENT-FILE in STUDENT-NO sequence otherwise (change to enrolment)
>
> For each SUBJECT-TO-BE-TAKEN
> > — insert the SUBJECT-NO into the ENROLLED-SUBJECT table
>
> For each SUBJECT-TO-BE-DROPPED
> > — remove the SUBJECT-NO from the ENROLLED-SUBJECT table

Validate subjects (process 2.2.1)

> For each SESSION
> > For each SUBJECT-TO-BE-TAKEN and SUBJECT-TO-BE-DROPPED,
> > > — validate input fields as per specifications
> >
> > If any errors are found
> > > — display an appropriate error message on the screen
> > > — ask for the subject to be entered again
> >
> > Find the SUBJECT-RECORD in the SUBJECT-FILE using SUBJECT-NO
> > If the SUBJECT-RECORD does not exist
> > > — display an appropriate error message on the screen
> > > — ask for the subject to be entered again
> >
> > For each SUBJECT-TO-BE-TAKEN
> > > — check that the STUDENT-NO does not exist in the ENROLLED-STUDENT table
> >
> > If the student is already enrolled for the subject
> > > — display an appropriate error message on the screen
> > > — ask for the SUBJECT-TO-BE-TAKEN to be entered again
> >
> > For each SUBJECT-TO-BE-DROPPED
> > > — check that the STUDENT-NO does exist in the ENROLLED-STUDENT table
> >
> > If the student is not enrolled for the subject
> > > — display an appropriate error message on the screen
> > > — ask for the SUBJECT-TO-BE DROPPED to be entered again

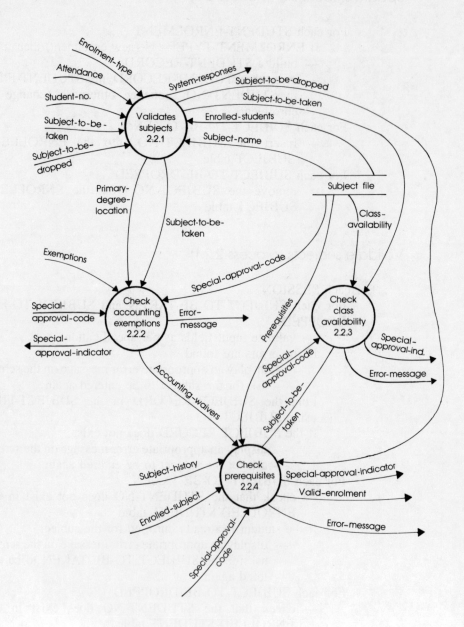

Figure 12—4   Validate enrolment (process 2.2)

For each SUBJECT-TO-BE-TAKEN and SUBJECT-TO-BE-DROPPED
    — display the SUBJECT-NAME on the terminal
    — ask if the subject is OK
    If no
        — ask for the subject to be entered again
Check that for each SUBJECT-TO-BE-DROPPED, a SUBJECT-TO-BE-TAKEN is entered
If ATTENDANCE = F (full-time) and ENROLMENT-TYPE = N (new enrolment)
    — allow enrolment in three subjects only per SESSION
If ATTENDANCE = P (part-time) and ENROLMENT-TYPE = N (new enrolment)
    — allow enrolment in two subjects only per SESSION

Check accounting exemptions (process 2.2.2)

If the SUBJECT-TO-BE-TAKEN is 14.970G or 14.996G or 14.973G and EXEMPTIONS include 14.940G and 14.941G and PRIMARY-DEGREE-LOCATION is 'A'
    then
        — display an appropriate error message on the terminal
        — ask for another SUBJECT-TO-BE-TAKEN to be entered
If the SUBJECT-TO-BE-TAKEN is 14.971G and PRIMARY-DEGREE-LOCATIONS is 'A'
    then
        — ask if special approval has been granted to take this subject
    If yes
        — (special approval has been granted)
        — ask for the SPECIAL-APPROVAL-CODE to be entered
        If the SPECIAL-APPROVAL-CODE is the same as the SPECIAL-APPROVAL-CODE from the SUBJECT-RECORD
            then
                — set the SPECIAL-APPROVAL-INDICATOR
            otherwise
                — (code not correct)
                — terminate the enrolment

If no
— (special approval not granted)
— ask for another SUBJECT-TO-BE-TAKEN
to be entered
If the SUBJECT-TO-BE-TAKEN is 14.907G or 14.996G and
EXEMPTIONS include 14.940G and 14.941G and PRIMARY-
DEGREE-LOCATION = 'O'
then
— display an appropriate error message on the terminal
— ask for another SUBJECT-TO-BE-TAKEN to be
entered
If the SUBJECT-TO-BE-TAKEN is 14.973G and PRIMARY-
DEGREE-LOCATION is 'O'
then
— check out special approval as detailed above
Set up ACCOUNTING-WAIVERS for the student

Check class availability (process 2.2.3)

For each SUBJECT-TO-BE-TAKEN
— check that the NO-STUDENTS-ENROLLED-IN-
CLASS is less than twenty
— and that the NO-STUDENTS-ENROLLED-IN-TUT
is less than fifteen
If not
— ask if special approval has been granted to exceed
class size
If yes
— ask for SPECIAL-APPROVAL-CODE to be
entered
If SPECIAL-APPROVAL-CODE does not match
with the SPECIAL-APPROVAL-CODE from
CLASS-AVAILABILITY
then
— terminate the enrolment
otherwise
— (special approval OK)
— set the SPECIAL-APPROVAL-
INDICATOR
If no
— (special approval has not been granted)
— ask for another SUBJECT-TO-BE-TAKEN
to be entered

Check prerequisites (process 2.2.4)

Assemble a full subject profile for the student from
— SUBJECT-HISTORY
— ENROLLED-SUBJECTS
— SUBJECT-TO-BE-DROPPED
— SUBJECT-TO-BE-TAKEN
— ACCOUNTING-WAIVERS
For each SUBJECT-TO-BE-TAKEN
— get the PREREQUISITES from the SUBJECT-FILE
— check that the PREREQUISITES are matched by the SUBJECT-PROFILE
If PREREQUISITES are not satisfied
— ask if special approval has been obtained to waive the prerequisites
If yes (i.e. special approval has been obtained) then
— ask for SPECIAL-APPROVAL-CODE
If the SPECIAL-APPROVAL-CODE is not equal to the SPECIAL-APPROVAL-CODE from the SUBJECT-RECORD
then
— terminate the entire enrolment
otherwise
— set   the   SPECIAL-APPROVAL-INDICATOR
If no, (i.e. special approval has not been obtained)
— ask for another SUBJECT-TO-BE-TAKEN to be entered

Extract reports (process 3)

Set out on the following pages are the detailed processes involved with producing the information reports from the system.

Extract special approval report (process 3.1)

For each STUDENT-RECORD in the STUDENT-FILE
If any SPECIAL-APPROVAL-SUBJECTS are present and have not yet been reported
then

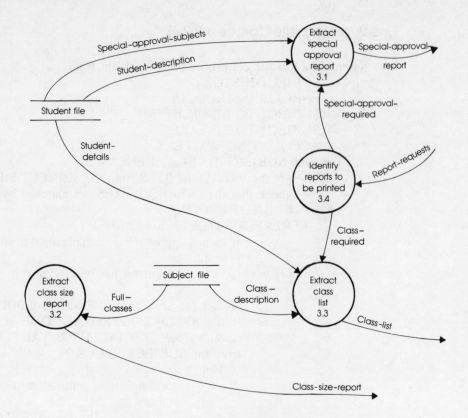

Figure 12—5 Extract reports (process 3)

— print the SPECIAL-APPROVAL-REPORT in the format shown in the specimen layout

Extract class size report (process 3.2)

For each SUBJECT, CLASS and TUTORIAL in the SUBJECT-FILE
— calculate the percentage full
If this percentage exceeds the parameter value
then
— extract the FULL-CLASSES data from the SUBJECT-FILE
— print a CLASS-SIZE-REPORT in the format shown in the specimen layout

Extract class list (process 3.3)

> If the CLASS-REQUIRED = ALL '9'
>> then
>>> — print all classes
>> otherwise
>>> — print only the class specified in CLASS-REQUIRED
> For each SUBJECT-RECORD to be printed
>> — extract the CLASS-DESCRIPTION from the SUBJECT-FILE
> For each ENROLLED-STUDENT
>> — extract the STUDENT-DETAILS from STUDENT-FILE
>> — print a CLASS-LIST in the format shown in the specimen layout

Identify reports to be printed (process 3.4)

This process interacts with the enrolling officer to determine which reports are to be printed off.

Data Dictionary

> ACCOUNTING-WAIVERS:
>> SUBJECT-NO*
> CHANGE-TO-ENROLMENT:
>> STUDENT-NO
>> SUBJECT-TO-BE-DROPPED:*
> CLASS:
>> CLASS-CODE
>> TIME-CLASS-OFFERED
> CLASS-AVAILABILITY:
>> SUBJECT-NO
>> SPECIAL-APPROVAL-CODE
>> SUBJECT-NAME
>> (SESSION
>> (CLASS:
>> NO-STUDENTS-ENROLLED-IN-CLASS)*
>> (TUTORIAL:
>> NO-STUDENTS-ENROLLED-IN-TUT)*)*

CLASS-DESCRIPTION:
    SUBJECT-NO
    SUBJECT-NAME
    CLASS:
    ENROLLED-STUDENT*
    TUTORIAL:
    TUT-ENROLLED-STUDENT*
CLASS-LIST:
    SUBJECT-DESCRIPTION
    CLASS:
    TUTORIAL:
    (STUDENT-DESCRIPTION
    SEX
    ATTENDANCE
    TELEPHONE-NO)*
CLASS-REQUIRED:
    SESSION
    SUBJECT-NO
    CLASS-TUT-CODE
CLASS-SIZE-REPORT:
    SUBJECT-DESCRIPTION
    SESSION
    (CLASS:
    NO-STUDENTS-ENROLLED-IN-CLASS
    (TUTORIAL:
    NO-STUDENTS-ENROLLED-IN-TUT)*)*
COMPLETE-STUDENT-HISTORY:
    STUDENT-NO
    STUDENT-HISTORY:
    SUBJECT-HISTORY:
ENROLLED-STUDENT:
    STUDENT-NO
ENROLLED-SUBJECT:
    SUBJECT-NO
ENROLMENT:
    STUDENT-ID:
    ENROLMENT-TYPE
    STUDENT-HISTORY:
    SUBJECT-TO-BE-TAKEN:*
    SUBJECT-TO-BE-DROPPED:*
    SUBJECT-HISTORY:
    [SPECIAL-APPROVAL-CODE]
EXEMPT-SUBJECT:
    SUBJECT-NO

EXEMPTIONS:
    EXEMPT-SUBJECT:*
FULL-CLASSES:
    CLASS-SIZE-REPORT
MCOM-SUBJECTS:
    UPDATE-TYPE
    CORE-UNIT-TYPE
    SUBJECT-DESCRIPTION
    SESSION
    CLASS-TUT-CODE
    PREREQUISITES:*
    TIME-CLASS-OFFERED
    SPECIAL-APPROVAL-CODE
NEW-ENROLMENT:
    STUDENT-NO
    SUBJECT-TO-BE-TAKEN:*
NO-STUDENTS-ENROLLED:
    ⟨NO-STUDENTS-ENROLLED-IN-CLASS
    NO-STUDENTS-ENROLLED-IN-TUT⟩
PRELIM-ENROL-FORM:
    STUDENT-DESCRIPTION
    SEX
    TELEPHONE-NO
    (SESSION
    SUBJECT-DESCRIPTION
    CLASS:
    TUTORIAL:)*
PREREQUISITE:
    SUBJECT-NO
PREREQUISITES:
    (SUBJECT-NO
    SPECIAL-APPROVAL-CODE)*
PRIMARY-DEGREE:
    ATTENDANCE
    PRIMARY-DEGREE-LOCATION
REPORT-REQUESTS:
    ⟨SPECIAL-APPROVAL-REQD
    CLASS-REQUIRED
    CLASS-FULL-PARAMETER⟩
SPECIAL-APPROVAL-REPORT:
    STUDENT-DESCRIPTION
    SUBJECT-NO*
SPECIAL-APPROVAL-SUBJECT:
    SUBJECT-NO

SPECIAL-APPROVALS:
    SPECIAL-APPROVAL-SUBJECT:*
STUDENT-DESCRIPTION:
    STUDENT-NO
    STUDENT-NAME
STUDENT-DETAILS:
    STUDENT-NO
    STUDENT-HISTORY:
STUDENT-ENROLMENT:
    [STUDENT-RECORD:]
    [⟨SUBJECT-TO-BE-TAKEN:
    SUBJECT-TO-BE-DROPPED:⟩]
STUDENT-ENROLMENT-HISTORY:
    EXEMPT-SUBJECT:*
    SPECIAL-APPROVAL-SUBJECT:*
    SUBJECT-ALREADY-PASSED:*
STUDENT-FILE:
    STUDENT-RECORD*
STUDENT-HISTORY:
    STUDENT-NAME
    SEX
    TELEPHONE-NO
    ATTENDANCE
    PRIMARY-DEGREE-LOCATION
STUDENT-ID:
    STUDENT-NO
    ENROLMENT-TYPE
STUDENT-PERSONAL-ID:
    STUDENT-ID:
    ATTENDANCE
STUDENT-RECORD:
    STUDENT-NO
    STUDENT-NAME
    SEX
    TELEPHONE-NO
    ATTENDANCE
    PRIMARY-DEGREE-LOCATION
    EXEMPT-SUBJECT:*
    SUBJECT-ALREADY-PASSED:*
    ENROLLED-SUBJECT:*
    SPECIAL-APPROVAL-INDICATOR
    SPECIAL-APPROVAL-REPORTED

STUDENT-REPORT-DATA:
    STUDENT-DESCRIPTION:
    [SPECIAL-APPROVALS]
STUDENT-UPDATES:
    ⟨STUDENT-RECORD:
    ENROLLED-SUBJECT:⟩
SUBJECT-ALREADY-PASSED:
    SUBJECT-NO
SUBJECT-DESCRIPTION:
    SUBJECT-NO
    SUBJECT-NAME
SUBJECT-DETAILS:
    SUBJECT-NO
    PREREQUISITES:
    CLASS-AVAILABILITY
SUBJECT-ENROLMENT:
    ⟨NEW-ENROLMENT:
    CHANGE-TO-ENROLMENT:⟩
SUBJECT-FILE:
    SUBJECT-RECORD:*
SUBJECT-HISTORY:
    EXEMPT-SUBJECT:*
    SUBJECT-ALREADY-PASSED:*
SUBJECT-RECORD:
    SESSION
    SUBJECT-NO
    SUBJECT-NAME
    PREREQUISITES:*
    CORE-UNIT-TYPE
    CLASS-DETAILS*
        CLASS-TUT-CODE
        SPECIAL-APPROVAL-CODE
        TIME-OFFERED
        NO-STUDENTS-ENROLLED
        STUDENT-NO:*
SUBJECT-REPORT-DATA:
    ⟨CLASS-DESCRIPTION
    SPECIAL-APPROVALS⟩
SUBJECT-TO-BE-DROPPED:
    SESSION
    SUBJECT-NO
    CLASS-TUT-CODE

```
SUBJECT-TO-BE-TAKEN:
    SESSION
    SUBJECT-NO
    CLASS-TUT-CODE
SUBJECT-UPDATES:
    STUDENT-NO
    SUBJECT-TO-BE-TAKEN:*
    SUBJECT-TO-BE-DROPPED:*
SYSTEM-RESPONSES:
    ERROR-MESSAGE*
    [SUBJECT-NAME]
TUT-ENROLLED-STUDENT:
    STUDENT-NO
TUTORIAL:
    TUT-CODE
    TIME-TUT-OFFERED
VALID-ENROLMENT:
    COMPLETE-STUDENT-HISTORY
    SUBJECT-TO-BE-TAKEN:*
    SUBJECT-TO-BE-DROPPED:*
    ENROLMENT-TYPE
VALID-MCOM-SUBJECTS:
    MCOM-SUBJECTS:
```

REFERENCES AND FURTHER READING

ACKOFF, R. L. 'Management Misinformation Systems', *Management Science*, December 1967, B147–B156.

ANTHONY, R. N. *Planning and Control Systems: A Framework for Analysis*, Division for Research, Graduate School of Business Admin., Harvard University, Cambridge, Mass., 1965.

ARGYRIS, C. 'Management Information Systems: the Challenge to Rationality and Emotionality', *Management Science*, February 1971, B275–B292.

BEOHM, B. 'Software Engineering — As It Is', *IEEE Fourth International Conference on Software Engineering*, September 1979, pp.11–21.

BJORN-ANDERSEN, N. (ed.), *The Human Side of Information Processing*, North Holland, Amsterdam, 1980.

BOSTROM, R. P. & J. S. HEINEN, 'MIS Problems and Failures: A Socio-Technical Perspective, Pt 1: The Causes', *MIS Quarterly*, September 1977, pp. 17–31.

BOSTROM, R. P. & J. S. HEINEN, 'MIS Problems and Failures: A Socio-Technical Perspective, Pt 2: The Application. Socio-Technical Theory', *MIS Quarterly*, December 1977, pp. 11–28.

BROOKES *et al. Information Systems Design*, Prentice-Hall of Australia, Sydney, 1982.

CARTER, D. M., H. C. GIBSON & R. A. RADEMACHER, 'A Study of Critical Factors in Management Information Systems for the US Air Force', *National Technical Information Service AD–A–009–647/9WA*, NTIS, Springfield, Va., 1975.

CHAPIN, N. 'New Format for Flowcharts', *Software-Practice and Experiences*, vol. 4, no. 4, February 1974, pp. 341–57.

CLIFTON, H. D. *Business Data Systems*, Prentice-Hall Inc., Englewood Cliffs, N.J., 1978.

DE MARCO, T. *Structured Analysis and Systems Specification*, Prentice-Hall Inc., Englewood Cliffs, N.J., 1980

ELSON, M. *Data Structures*, Science Research Associates, Chicago, Ill., 1975.

GANE, C. & T. SARSON, *Structured Systems Analysis*, Prentice-Hall Inc., Englewood-Cliffs, N.J., 1979.

GOTLIEB, C.C. & L.R. GOTLIEB, *Data Types and Structures*, Prentice-Hall Inc., Englewood Cliffs, N.J., 1978.

GROUSE, P. J. 'Flowblocks — A Technique for Structured Programming', *ACM SIGPLAN Notices*, vol. 13, no. 2, February 1978, pp. 46–56.

JACKSON, M. A. *Principles of Program Design*, Academic Press, London, 1975.

LUCAS, H. *Towards Creative Systems Design*, Columbia University Press, New York, 1975a.

LUCAS, H. *Why Information Systems Fail*, Columbia University Press, New York, 1975b.

MINTZBERG, H. 'The Manager's Job: Folklore and Fact', *Harvard Business Review*, July/August 1975, pp. 49–61.

MUMFORD, E, F. LAND & J. HAWGOOD, 'A Participative Approach to the Design of Computer Systems', *Impact of Science on Society*, vol. 28, no. 3, 1978. pp. 235–53.

NASSI, I. & B. SHNEIDERMAN, 'Flowchart Techniques for Structured Programming', *ACM SIGPLAN Notices*, vol. 8, no. 8, August 1973, p. 12.

YOURDON, E. & L. CONSTANTINE, *Structured Design*, Prentice-Hall Inc., Englewood Cliffs, N.J., 1979.

INDEX